DESIGN OF WELDED STEEL STRUCTURES
PRINCIPLES AND PRACTICE

DESIGN OF WELDED STEEL STRUCTURES
PRINCIPLES AND PRACTICE

Utpal K. Ghosh

CRC Press
Taylor & Francis Group
Boca Raton London New York

CRC Press is an imprint of the
Taylor & Francis Group, an **informa** business

CRC Press
Taylor & Francis Group
6000 Broken Sound Parkway NW, Suite 300
Boca Raton, FL 33487-2742

First issued in paperback 2017

© 2016 by Taylor & Francis Group, LLC
CRC Press is an imprint of Taylor & Francis Group, an Informa business

No claim to original U.S. Government works

ISBN-13: 978-1-4987-0801-2 (hbk)
ISBN-13: 978-1-138-74875-0 (pbk)

Visit the Taylor & Francis Web site at
http://www.taylorandfrancis.com

and the CRC Press Web site at
http://www.crcpress.com

To Manjula, Supriti, and Indranil.

Contents

Preface

Arc welding represents the state of the art for fabrication of steel structures. In this technique, controlled heat is applied for joining the components of structures along the line of connection. The performance of welded steel structures depends on a number of factors, of which characteristics and quality of welded joints are the most important. In recent decades, there has been phenomenal development in this technology, which has made arc welding highly attractive in the construction industry globally. In fact, welded construction has already proved to be of great advantage to stakeholders, namely, architects, structural engineers, contractors, and their clients—end users. It is necessary that more people acquire knowledge and experience in this field to make welding a more powerful tool for an expanding construction industry. It is in the backdrop of this situation that the present book is immensely relevant.

The book deals with both the principles and practice of welding technology, which is required for satisfactory design of welded steel structures and thus should be of deep interest and value not only to the practicing engineers in the design office or in the workshop, but also to teachers and students in academia.

The presentation of text in this book is somewhat different from that in normal engineering books. This book should be regarded as a complementary work to the more analytical studies, which present worked-out examples. Consequently, topics not usually covered in existing textbooks but are nevertheless important for the understanding of the subject have found place in this work.

The text can be broadly divided into four parts. Chapters 1 through 6 deal with the basics of arc welding and include brief notes on the salient features of modern arc welding processes, types and characteristics of welded joints, their common defects and recommended remedial measures, and quality control aspects in the workshop. Chapters 7 through 9 primarily deal with analysis and detail design of welded structures. Chapters 10 through 15 provide useful information and discussions on the detail design of joints in respect of some common welded steel structures. The concluding chapter (Chapter 16) discusses cost factors involved in welded steelwork.

The material covered in the text has been drawn from the vast pool of accumulated knowledge and experience of distinguished engineers gained through studies in different countries, primarily Europe and America, and supplemented by the author's own experience. As far as possible, references to the published literature have been mentioned at the end of each chapter. The author thankfully acknowledges his indebtedness to these writers. However, if the ideas of earlier writers have appeared in the book without appropriate acknowledgment, it is quite unintentional, and the author would like to extend his apologies. If such instances are brought to the author's notice, the same will be gratefully acknowledged in the subsequent edition of the book.

The author has also gained immense knowledge from personal interactions with a large number of individuals. Many a time, small points raised in discussions have led to a major change in the text or the inclusion of an additional topic. It is practically not possible to list such individual names. However, the author gratefully acknowledges his debt to each of them. Special thanks are due to the author's longtime colleague and friend Amitabha Ghoshal for his support throughout the preparation of the manuscript and valuable suggestions. The author thanks his son, Indranil, and daughter-in-law, Supriti, for their assistance in the preparation of the manuscript, particularly during his several visits to the United States. Thanks are due to Tilokesh Mallick for the long hours he spent ungrudgingly for keying in the bulk of the text into a computer and helping the author in surfing the Internet as well as other various activities. Sanjoy Bera also deserves special mention for drawing electronically all the figures appearing in the book.

And last, but by no means the least, the author is grateful to his wife, Manjula, for her encouragement and support in writing the book.

Utpal K. Ghosh

Author

Utpal K. Ghosh worked, among others, with Freeman Fox and Partners, London; Sir William Arrol & Co. Ltd., Glasgow, Scotland; and Braithwaite Burn & Jessop Construction Co. Ltd., Kolkata, India, after graduating in civil engineering in 1954 from Bengal Engineering College, Shibpur, Calcutta University (currently Indian Institute of Engineering Science and Technology). Subsequently, he set up his own practice as a consulting engineer.

During his long career, he participated in the planning, design, fabrication, erection, and overall management of a wide variety of projects, such as bridges and industrial structures, which included new construction as well as repair and rehabilitation work. He has worked on projects in several countries, including the United Kingdom, New Zealand, Malaysia, Indonesia, Singapore, and India.

He has published a number of articles and is the author of two books entitled *Design and Construction of Steel Bridges* and *Repair and Rehabilitation of Steel Bridges*.

He is a Chartered Engineer and is a Fellow of the Institution of Engineers (India), a Member of the Institution of Civil Engineers (UK), and a Member of the Institution of Structural Engineers (UK).

1

Electric Arc Welding Processes

ABSTRACT The chapter begins with the basic principles of making a welded joint and goes on to briefly describe the development of welding processes, from 1400 BC to modern times. The arc welding processes commonly used now are described along with their advantages and disadvantages. Criteria for the selection of a particular process for welding are discussed. The chapter concludes with a short discussion on the safety aspects to be considered for using any of these welding processes.

1.1 Introduction

Welding has now become the most common method of fabrication of steel structures in preference to traditional techniques such as riveting and bolting. Consequently, for a designer, a basic knowledge of the welding process is necessary in order to properly design welded structural steelwork, particularly the design of joints.

Welding is the process of joining two metal components by bringing them to the molten state at the faces to be joined and then allowing the molten metal to intermingle and, when cool, establish a metallurgical bond between the components. Thus, the process is essentially a fusion process.

The technology of forge welding is believed to have been used first by the Syrians in about 1400 BC. Since then, the technology may have been lost and rediscovered a number of times by our ancient forefathers residing in various parts of the globe. Coming to the more recent past, the term *welding* has been generally associated with the village blacksmith's smithy shop, where two metal pieces are softened in the concentrated heat developed by charcoal fire and joined together, or *welded*. In a modern structural fabrication shop, however, the welding process commonly used is the electric arc process. This concept of using electric arc as a suitable source of intense heat to reduce the metal into a liquid was first used in a practical application in the nineteenth century, by forming an electric arc between a carbon piece and the metal workpiece (UK Patent No. 12984 of 1885, Benardos and Olszewski), hence the name *electric arc welding* process. Subsequently, the carbon piece was replaced by a steel rod, called *electrode*.

Essentially, the arc is a sustained spark formed between the workpieces to be welded and the electrode. As the electric arc is brought close to these workpieces, a low-voltage (15–35 V), high-current (50–1,000 A) electric arc is formed between the tip of the electrode and the work, and the temperature at the location under the tip of the electrode jumps to approximately 6,500°F (3,600°C). This concentrated heat melts metal from each component as well as the electrode, forming a common pool of molten metal, called the *crater*. This pool on cooling forms a solid bond between the components, thereby providing a continuity of metal at the interface. By moving the electrode along the line of the joint, the surfaces to be joined are welded together along their entire length. Normally, the composition of the electrode is so chosen that the resultant weld becomes stronger than the connected components.

During the past four decades, welding technology has undergone phenomenal developments, and quite a number of welding processes are now available for use in the fabrication industry. In all these processes, the arc is shielded by a number of techniques.

The most common of these techniques are a chemical coating on the electrode rod/wire, inert gases, and granular flux compounds. The primary purposes of such shielding are to

- Protect the molten metal from the effects of air
- Add ingredients for alloying the resultant weld metal
- Control the melting of the electrode and thereby ensure effective use of the arc energy

Some of the common processes used in fabrication of welded steel structures are briefly discussed in the following paragraphs.

1.2 Manual Metal Arc Welding

The manual metal arc (MMA) welding process is one of the oldest processes of arc welding and is also known as *stick electrode welding*, *shielded metal arc welding*, and *electric arc welding*. In this process, the electrode is a steel stick coated with flux containing alloying elements such as manganese and silicon. The electrode stick is generally 350 mm long. The diameter of its steel core is 3.2–6.0 mm to match the level of the current used. The bare section of the electrode is clamped to an electrode holder, which is connected to the power source by a welding cable. The holder is held by hand. As an arc is initiated between the end of the handheld electrode and the parent metals in the location of the joint line by touching the electrode tip against the work and then slightly withdrawing it, the heat melts both the parent metals and the electrode, forming a pool of molten weld. The mechanical properties of the weld metal are somewhat improved due to the presence of the molten flux layer and the gas generated by the flux, which also protects the molten metal from oxidation. The welder moves the electrode forward with a uniform travel speed to achieve continuity of the weld deposit and, at the same time, moves it toward the weld pool to keep the arc gap at a constant distance. These two simultaneous movements are necessary to produce a uniformly wide and continuous weld deposit. The skill of the welder in keeping a constant feed rate as well as travel speed is thus an important factor to achieve a good quality of MMA welding. Apart from this, the quality of MMA welding is also dependent on certain other factors, such as the compatibility of the electrode, as well as the welding parameters (voltage, current, etc.).

The selection of electrodes depends largely on the application in question, such as joint detail, service requirement, or weld position. Service requirement may dictate use of low-hydrogen electrodes for structures subjected to impact or dynamic loadings, such as cranes or bridges. Low-hydrogen electrodes are also recommended for welding thicker plates. It must also be ensured that the welders are familiar with the use of such special categories of electrodes. For this purpose, qualification tests for welders and approval of welding procedures are recommended, as discussed in Chapter 6.

An electrode gradually gets reduced in length as it is melted and needs replacement when the length is about 50 mm. At this stage, the arc is extinguished, and the solidified flux, often called *slag*, deposited on the surface of the solidified weld, is carefully removed, before a fresh electrode is used to continue the welding. A schematic diagram of MMA welding process is shown in Figure 1.1. Because of the limited length of the electrodes in the MMA welding process, it is not possible for the welder to deposit continuous welds beyond a certain length, depending on the diameter of the electrode. Thus, this process presents some problems where a continuous weld is preferred. However, despite this drawback, MMA has gained immense popularity over the years among the structural fabricators for a number of advantages it offers. The most important among these advantages is its versatility of applications, the simplicity of the equipment, and its lightness (can be moved to almost any location, both within the workshop and in the construction site). The system can also be operated for any position of the welding (downhand, horizontal, or overhead) without any problem. It has a lead of about 20 m from the power supply point, making it immensely useful particularly at work sites. Furthermore, the capital cost of a unit is comparatively low, which makes it quite attractive from economic point of view as well.

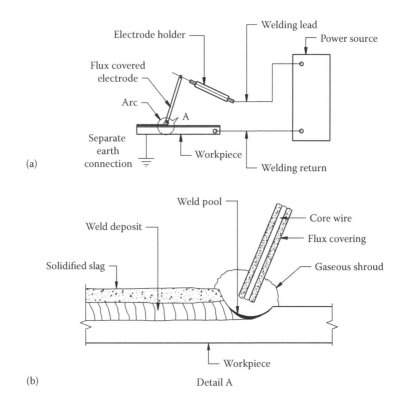

FIGURE 1.1
(a, b) Schematic arrangement of manual metal arc (MMA) welding process.

1.3 Metal-Active Gas Welding

The metal-active gas (MAG) welding process is also known as *metal inert gas welding, gas metal arc welding, metal arc gas-shielded welding*, or *CO₂ welding*. This is also a manually operated system but has the option to be used with a mechanical traversing device, thereby making it a semiautomatic welding process. In this process, the electrode normally consists of bare solid metal wire of 0.9–1.6 mm diameter containing alloying elements and is fed at a constant speed from a spool by a motorized unit. An inert gas that does not react with the molten steel is used to shield or protect the arc and the molten weld metal. The shielding gases include carbon dioxide or a mixture of argon and carbon dioxide/oxygen. The selection of the gas to be used depends on its compatibility with the properties of the components being welded, the joint type, the metal thickness, as well as the mode of the welding operation.

The current is determined by the preset speed of the wire feed, and the arc length is determined by the preset power supply unit. In this semiautomatic process, the welder has to control these different parameters as well as maintain a constant height of the nozzle above the weld pool. A schematic diagram of the process is shown in Figure 1.2. Semiautomatic gas-shielded welding with bare wire has an advantage over MMA welding in that it does not require changing of electrodes, and therefore, it is a continuous process, requiring no de-slagging between successive runs. Consequently, it is comparatively a faster process, leading to reduction in the fabrication costs. However, it requires a certain degree of sophistication in its execution, needing trained skilled welders to produce good-quality welds. Also, it is best

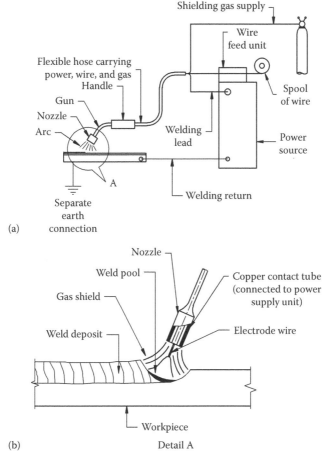

FIGURE 1.2
(a, b) Schematic arrangement of metal-active gas (MAG) welding process.

operated in the closed and controlled environment of a workshop, rather than the open site environment where high-wind conditions may impede the effectiveness of the gas shield. In a gas-shielded process, a nozzle is used to feed the wire and the gas on the weld location. Thus, it is necessary to provide sufficient space for the nozzle to access the weld area at the required angle. This limitation may need re-detailing of the joint. Alternatively, MMA process has to be used for such awkward locations.

In this process, normally flux is not necessary. However, in some special applications, a tubular wire containing flux is used. The flux, enclosed in the tubular wire, produces gas on burning, which shields the molten metal. Some of the advantages of this variation in the system are as follows:

- High-quality weld at lower cost with less effort by the welder
- Better weld profile, for example, the fillet welds in the horizontal or vertical positions
- Relatively high travel speed and high deposition rate
- Considerably reduced spatter
- Reduced distortion
- Easy to weld

1.4 Submerged Arc Welding

The submerged arc welding (SAW) is a high productivity mechanized process in which the arc is completely *submerged* under a covering layer of fusible granular flux (and hence the name, see Figure 1.3). A bare electrode wire of 2.4–6 mm diameter, coiled in a spool, is fed continuously into the weld zone by mechanically operated drive rolls. As an arc is struck between the end of the electrode wire and the work, the flux is simultaneously deposited continuously from a hopper, around the weld on the surface of the joint. The molten weld pool and the arc zone are protected from contaminants in the atmosphere as these are covered under the blanket of the flux. The conductive molten flux provides a current path between the electrode and the work. The voltage and the current are controlled automatically at a preselected value.

SAW is commonly used as a fully automatic process in which the work is kept stationary and the electrode and the drive assembly are traversed along the joint by mechanical drive system. Alternatively, the work can also be moved in relation to the welding head. The electrode feed rate is so adjusted that the preset arc length remains constant throughout the operation. SAW can also be used in semiautomatic mode, depending on the nature and extent of the work. For this mode, special hand-operated guns are available.

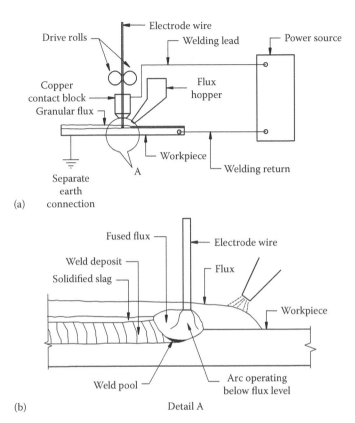

FIGURE 1.3
(a, b) Schematic arrangement of submerged arc welding (SAW) process.

The arc melts a portion of the flux that intermixes with the molten weld from the electrode and the work and forms a glass-like slag that is lighter in weight than the deposited weld metal and floats on the surface covering the weld as it cools.

Different varieties of electrodes can be used. These can be single or multiple solid wires, tubular wires, or strips. The selection of the electrode and the flux should be made with extreme caution, because the chemistry and properties of the weld metal will vary based on the electrode and the flux selected. The manufacturers' recommendations in this respect become very important as these consumables must also match the base metals being welded.

The SAW process has the advantages of high metal weld quality and smooth and uniform weld finish. It is particularly useful for welding long joints. The use of a continuous wire feed coupled with a mechanized operation produces a high weld deposition rate, thereby enhancing the productivity rate considerably. Also, with good process design and control, deep weld penetration is achieved. The effective shielding provided by the flux produces sound, uniform, ductile, corrosion-resistant welds having good impact value. Additionally, the working conditions are vastly improved compared to other processes, because the flux hides the arc and consequently the welding fume is minimal; also, there is hardly any spatter of the weld due to the blanket of flux over the weld. SAW process practically needs no edge preparation. Single-run welds can be made in thick plates. Because of these advantages, SAW process is used extensively in the structural workshops the world over. However, the process is limited to welding in the flat position only.

1.5 Stud Welding

Welded steel stud shear connectors are commonly used in steel-concrete composite girders. A mechanized device or *gun* is used for welding the bottom ends of the studs on the top surface of the top flanges of steel girders. The steel stud, which typically has a head on the top, acts as an electrode and is held in a chuck of the gun, which is connected to the power supply. The stud, with a ceramic ferrule around it, is pressed firmly on the surface of the steel by means of the gun. As soon as the trigger in the gun is pressed, the current is switched on and the stud is moved away automatically from the steel surface of the girder to form an arc. The resulting heat melts the end of the stud and the top surface of the steel plate of the girder to form a molten pool. The stud is then plunged into the pool, which produces a fillet weld around the stud. The current is then automatically cut off.

For producing good-quality welds, it is necessary to correctly establish the various parameters, such as current, voltage, arc time, and force. Barring this, the operation of the equipment is relatively simple and offers an accurate and fast method of welding shear connectors on to the girders. Usually, the stud welding gun welds one stud at a time. For major works where large number of studs are required to be welded, multistud setups are available for use. Normally, stud welding is done using direct current.

1.6 Control of Welding Parameters

In order to achieve the required strength in any welded joint, the bonding between the weld metal and the connecting components must be of acceptable quality throughout the length of the joint. This implies that the heat input by the arc must be maintained at the optimum level throughout the operation. Heat input, in turn, depends on the following:

- Arc voltage
- Arc current
- Travel speed
- Arc length
- Electrode feed rate

Thus, it is imperative that these parameters are controlled to achieve a good quality of continuous weld.

1.7 Selection Criteria of Welding Process

While costs mostly dictate the selection of a particular process for welding, there are other factors also, which need due consideration. Thus, a holistic approach is necessary while choosing the appropriate process for a particular application. Some of these topics are discussed hereunder.

1.7.1 Costs

Costs in an arc welding operation include the following:

- Direct (welding) labor
- Indirect (associated) labor
- Cost of consumables (electrode, gas, flux, etc.)
- Equipment capital cost
- Equipment maintenance cost
- Overhead expenses

In addition to the welder time spent on actual welding (i.e., arcing time), direct labor costs should include the idle time while the electrode is being changed (for MMA welding) or aligning the joint (in SAW or MAG welding) and also waiting time between successive operations. The arcing time expressed as a percentage of the total time is termed *duty cycle*. Thus,

$$\text{Duty cycle} = \left(\frac{\text{Arcing time}}{\text{Total time}} \right) \times 100 \tag{1.1}$$

Obviously, the process with the maximum duty cycle would appear to be the most attractive one for selection. Consequently, for long joints, SAW and

MAG processes offer best solution as time would be low, leading to faster operation and increased productivity. However, the scenario is reversed for joints with short weld run and at different locations. In such cases, MMA has an edge, in spite of its high idle time.

1.7.2 Location of the Work

Location of the work is an important factor in selecting the welding process. Controlled and protected environment of the fabrication shop ideally suits both SAW and MAG welding processes. Regular fabrication shops are mostly equipped with the infrastructure required for these processes, such as cutting tools, jigs, fixtures, and manipulators (positioners). These facilities contribute largely to the production of welded structures not only of acceptable quality but also in cost-effective quantity. These facilities may not be available in the construction site, where MMA process may be well suited for fabrication in open and unprotected environment.

1.7.3 Welding Position

Mechanized processes such as SAW and MAG welding are eminently suitable for continuous and downhand welding. For vertical or overhead welding, these mechanized systems are not suitable, unless manipulators are used to rotate the workpiece suitably to enable the welding head to perform downhand welding. At work sites, these facilities may not be available. In such cases, MMA welding process provides an ideal solution for the obligatory overhead welding.

1.7.4 Access

Ease of access for electrode, welding gun, or welding head also needs to be examined while selecting a particular system.

1.7.5 Composition of Steel

Composition of steel may make a particular component susceptible to crack, needing special procedures to reduce or eliminate the risk of such crack formation. Hydrogen content of the weld metal can be better controlled in SAW and MAG processes than in MMA process. Issues related to risk of crack formation will be discussed in Chapter 3.

1.7.6 Availability of Welding Consumables

Generally, for MMA welding process, electrodes of various diameters, and with a variety of flux combinations to suit a wide range of applications, are easily available in the market. In this respect, MMA process has an edge over other systems.

1.7.7 Availability of Skilled Welders

Welding work, being a sophisticated process, needs trained and reasonably skilled workmen. Therefore, for any process to be adopted for the welding work, it has to be ensured beforehand that workmen with sufficient skill and experience are easily available in the location of the workplace. Otherwise, skilled welders may have to be imported from other locations. This problem is generally faced where SAW and MAG processes are contemplated; for MMA process, this problem may not be so acute.

1.8 Safety Aspects

Welding work is generally fraught with danger. However, with proper precautions and protections, the risks of injury or fatality can be substantially reduced. Common welding processes involve electric arc, which may lead to injury from burns due to extreme heat, flames, or sparks. To avoid this risk, it is necessary for the welders to wear protective clothing, such as heavy leather gloves and protective long-sleeve jackets. Furthermore, the extreme brightness of the weld area may cause serious damage to the eyes of the welder. To prevent this exposure, goggles and helmets with dark face are to be worn. To protect nearby workers and bystanders from the exposure to the UV light from the arc, transparent welding curtains, made of special plastic material, are often used to surround the weld area.

Also, welders are often subjected to exposure to various gases (e.g., carbon dioxide) and particulate matter, which may prove to be dangerous. It is therefore imperative that adequate arrangement for ventilation is provided in the working area.

Bibliography

1. Gourd, L.M., 1995, *Principles of Welding Technology*, Edward Arnold, London.
2. Houldcroft, P., and John, R., 1988, *Welding and Cutting*, Woodhead-Faulkner, Cambridge.

2

Welded Joints

ABSTRACT   The chapter describes the various types of welds and welded joints that are commonly used in steel fabrication and provides an idea about the characteristics of these joints—the weld as well as the metal around them (heat-affected zone). The interacting variables that influence the properties of welded joints are explained. The chapter ends with a brief discussion on the residual stresses that are locked up during cooling of a welded joint.

2.1 Introduction

It is now generally acknowledged that arc welding can be utilized in most types of steelwork fabrication more efficiently, and often more economically, than was possible with riveting system. In order that these advantages are maximized, it is necessary that the structure should be designed *for* welding. In other words, care must be taken at all stages of design to give due consideration to the stress-carrying capacities of the different types of welded joints, their behavior under different load conditions, and the basic features of the welding methodology to be employed to produce the joints. Integral to these variables is the design of the welds, whose type, size, and position become important to produce the most effective welded joints.

This chapter introduces the various types of welds and welded joints employed in fabricating welded structures and deliberates on the characteristics of these joints.

2.2 Types of Welds

In structural fabrication work, two types of welds are commonly used to form welded joints. These types are discussed in the following paragraphs.

2.2.1 Fillet Weld

In fillet weld, the weld metal is deposited outside the profile of the joining elements. Figure 2.1 shows a few typical fillet welds.

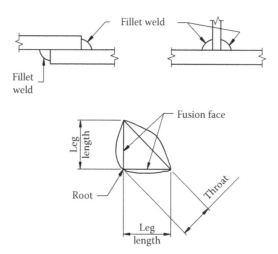

FIGURE 2.1
Fillet weld.

2.2.2 Butt Weld

In butt weld, the edges of the members are butted against each other and joined by fusing the metal to produce a continuous joint. Thus, butt weld is made within the surface profile of the joining members. Depending on the current used, the arc can melt the metal to a certain depth only. If the thickness of the members being joined is more than this depth, the edges of the members are required to be *prepared* to form a groove along the joint line, so

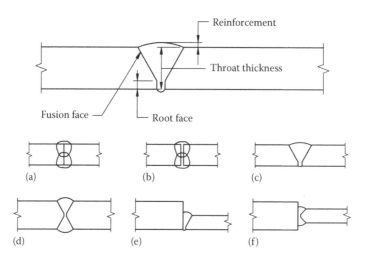

FIGURE 2.2
Butt weld: (a) close square butt, (b) open square butt, (c) single-V butt, (d) double-V butt, (e) single-bevel butt, and (f) double-bevel butt.

that the continuity of the joint through the full thickness is achieved. The prepared groove is then filled by weld metal from the electrode. Figure 2.2 shows a few common types of butt welds.

Properties of the parent metals, edge preparation, selection of the electrode, and welding parameters (current, speed, voltage, etc.) play a vital role in developing the strength of a butt welded joint.

2.3 Types of Welded Joints

A variety of joints can be formed by using the two types of welds, namely, fillet weld and butt weld. These joints can be made up from the four basic configurations described in the following paragraphs.

2.3.1 Butt Joints

Butt joints are commonly used to join lengths of plates (as in plate girders).

2.3.2 Tee Joints

Depending on the service requirement, tee joints can be formed either by fillet weld or by butt weld. Typical examples of these joints are at flange to web connections, stiffeners welded to the web, and flanges of plate girders. Figure 2.3 shows some examples of such fillet welded joints.

2.3.3 Corner Joints

Corner joints are commonly used in box girders and can be formed either by fillet weld or by butt weld, as shown in Figure 2.4.

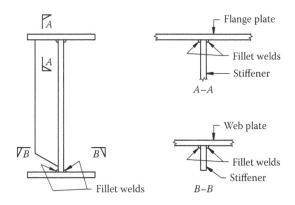

FIGURE 2.3
Tee joint.

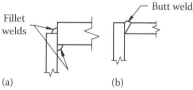

FIGURE 2.4
Corner joint: (a) by fillet welds and (b) by butt weld. (a) (b)

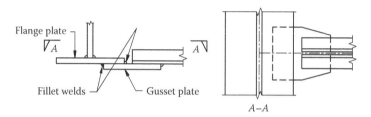

FIGURE 2.5
Lap joint.

2.3.4 Lap Joints

Lap joints are commonly formed by fillet welds. Bonding in lap joints are at the interfaces of the fillet welds only. These joints are widely used in structural fabrication work, such as connecting gussets to main members, as shown in Figure 2.5.

2.4 Heat-Affected Zone

The weld metal that joins the two steel components is essentially a mixture of the molten steel from each component and also from the electrode. It is a common practice to make this metal stronger than the components by judicious adoption of the composition of the electrode. As the metal of the molten weld pool cools, only a small amount of heat escapes through the surface of the weld pool, leaving the majority of the heat to flow through the parent metal on either side of the joint (see Figure 2.6). As a result, steel up to a certain distance is subjected to a severe thermal cycle—heating and then rapidly cooling. This causes change in the microstructure and in the properties of the steel in the region. This region of the parent material is called *heat-affected zone* (HAZ). The parent metal near the fusion boundary is subjected to a maximum temperature close to the melting point of steel. This temperature gradually falls as the distance from the fusion boundary increases until

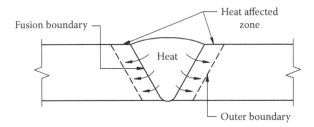

FIGURE 2.6
Heat-affected zone.

the outer boundary of HAZ is reached where the temperature is below the range for metallurgical change.

With the progressive cooling of the molten metal, the metallurgical structure of the steel in HAZ is changed from a ductile to a hard form. When the hardness is above a critical level, the metal becomes prone to cracks.

Two factors primarily contribute to the level of hardness in the HAZ, namely:

- Chemical composition of steel
- Rate of cooling

2.4.1 Chemical Composition of Steel

Carbon is primarily the strengthening element in steel. However, increased carbon content impairs ductility and weldability of the steel. Therefore, to attain better physical properties in steel (while keeping the carbon level low), other admixtures or alloys are generally added during the process of steel making. Some of these alloys (e.g., manganese and chromium), however, increase the hardness of the metal and consequently the risk of crack in the HAZ.

The relative influence of chemical contents on the weldability of a particular steel is guided by the value of *carbon equivalent* (CE), which is derived from the following empirical formula:

$$CE = C + \frac{Mn}{6} + \frac{Cr + Mo + V}{5} + \frac{Ni + Cu}{15} \qquad (2.1)$$

In Equation 2.1, the chemical symbols represent the percentage of the respective elements in the steel. It may be noted that the amount of each alloying element is factored according to its likely contribution toward hardening of the steel. Several other variants of this formula are available and are recommended by different authorities. This formula has been adopted by the International Institute of Welding and is widely followed in different countries.

In structural steel, the value of CE ranges from 0.35 to 0.53. As a general rule, however, a particular steel is considered weldable when the CE is less than 0.4. With increased value of CE, use of low-hydrogen electrodes and preheating of the components to be joined become important.

The relationship between the cooling rate and CE has special significance for cracking. Faster cooling rate for material with low CE may be tolerable as the risk of cracking is less. Thus, the higher the CE, the lower will be the tolerable cooling rate and consequently the harder and more brittle will be the HAZ and more susceptible will it be to cracking. The level of hardness in steel is measured by Vickers Pyramid Number (VPN). In this system, an indenter is forced into the surface of the steel and the size of the impression is compared with the preset standard. VPN for steel to be used for fabrication should ideally be in the range of 190–200. This should be taken into account in the design stage while specifying the steel grade to be used. In this context, reference may be made to Chapter 6.

2.4.2 Rate of Cooling

The moving heat source used in arc welding process induces steep temperature gradients around melt zone. Figure 2.7 illustrates a typical isotherm and temperature gradients along and across the line of movement of the arc. It may be particularly noted that the movement of the arc results in the piling up of the isotherms at the leading edge.

The direction of the flow of the heat from the source area depends on the thickness of the plates. In the case of thick plates, the heat flows in the horizontal as well as vertical direction, while in the case of thin plates heat flow is horizontal. The phenomenon is illustrated in Figure 2.8.

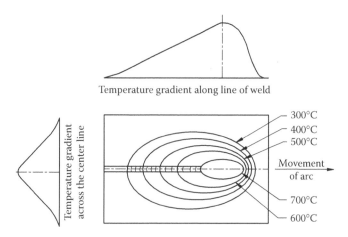

FIGURE 2.7
Isotherms during welding on steel plates.

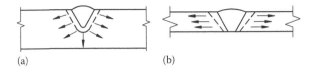

FIGURE 2.8
Heat flow: (a) thick plate and (b) thin plate.

Rapid cooling makes steel comparatively hard and brittle and consequently prone to cracking. Rate of cooling in the HAZ depends on a number of factors:

- High heat input entails a slower cooling rate and thereby reduces the risk of cracking in the HAZ.
- A thick section cools more rapidly than a thin section and is thus more susceptible to cracks in the HAZ.

It is a common practice to preheat thick components to reduce the temperature gradient. This reduces the level of hardening of the metal and also the chance of consequent cracks in the HAZ. Preheating is also useful in dispersing hydrogen from the weld metal and thereby reducing the risk of its embrittlement. Other uses of preheating are removal of surface moisture in humid conditions and also maintaining the ambient temperature in a cold environment. This treatment should, however, be used only selectively, because it entails additional expense.

2.5 Interacting Variables

The final properties of a welded joint may be influenced by any one or more of the following interacting variables involved in the welding process.

2.5.1 Composition of the Parent Metal, Electrode, and Flux

The weld metal consists of a mixture of materials obtained from the parent metal, the electrode, and the flux. All these materials have important effects on the final properties of the solidified weld.

2.5.2 Welding Process

Welding process determines the size of the weld pool and its geometry.

2.5.3 Environment

During the welding process, moisture and gases such as oxygen, nitrogen, and hydrogen are likely to penetrate into the weld pool. Of these gases, hydrogen makes the weld prone to cracks.

2.5.4 Speed of Welding

Speed of welding has a direct influence on the solidification speed of the weld, which affects the properties of the final weld.

2.5.5 Thermal Cycle of Weld

The pattern of thermal cycle also affects the properties of the weld metal.

2.5.6 Size and Type of Joint

In the case of thick plates, multirun welds are preferable as these are likely to reduce residual stress. This entails consumption of more time leading to costlier (but better) joints.

2.5.7 Manipulation of Electrodes

In order to produce good weld, it is necessary to manipulate the electrodes properly so as to avoid craters at the edge of the section, which may lead to cracks in future.

2.6 Residual Stresses

During cooling, the weld metal undergoes plastic deformation. As the weld cools and the plastic deformation is completed, there will be some stresses locked up in the joint. These stresses are termed *residual stresses*. This stress pattern in a butt welded joint is shown in Figure 2.9. It will be noted that moving out from the centerline of the joint, the residual tension is reduced to zero, and then, there is a zone of compression. This residual stress pattern is present in most of the welded joints.

Residual stress in large fillet welds on thick plates is illustrated in Figure 2.10. In this case, the two thick components are in contact prior to welding. As the welds cool, as there is no possibility of plate movement, transverse shrinkage stress (tension) is developed in the weld, and in case of severe restraint, the weld becomes prone to cracking. A small gap between the plates, prior to welding, would reduce the proneness to crack.

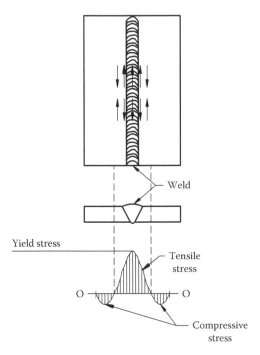

FIGURE 2.9
Residual stresses in butt weld.

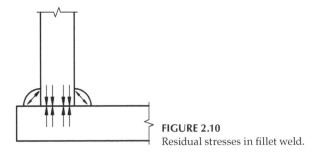

FIGURE 2.10
Residual stresses in fillet weld.

The presence of residual stresses in many cases (such as in low-pressure pipe work and storage tank) does not present appreciable problems in the behavior of the joints during service conditions. Consequently, joints in such cases are not required to undergo any postweld treatment for relieving the residual stresses. However, in certain situations it may be necessary to consider the effects of residual stresses. For example, the presence of residual stresses in very cold environment may embrittle the material, causing brittle fracture (Chapter 5). Also fluctuating loading pattern during service conditions (as in a crane girder) may lead to fatigue cracks due to the presence of residual stresses. Effects of fatigue in welded joints have been discussed in Chapter 9. Furthermore, in certain types of environment, some steel metals are susceptible to corrosion in the tension zone, and the presence of residual tension stress in the welded joints located in this zone enhances the risk of cracking due to stress corrosion. In such cases, it is advisable to subject the welded joints to stress-relieving treatment.

2.6.1 Heat Treatment

The most common method of stress relieving a welded joint is by heat treatment. With the application of heat, the yield stress of steel decreases. Thus, if a welded joint is so heated that its temperature reduces the yield stress to a value that is below the residual tensile stress, a localized plastic deformation will occur, and the tensile stresses will be reduced. Simultaneously, the compressive stresses will also be reduced to balance the equilibrium of the heated zone. The approximate temperature for effective stress relieving depends on the relationship between the temperature and the type of steel (alloy content) on which depends the corresponding yield stress.

A note of caution needs to be highlighted here. As thermal treatment involves localized heating, there is always a danger of differential expansion and contraction causing new residual stresses. Therefore, heating and cooling need to be carefully controlled. This can be achieved by maintaining the temperature distribution between the weld centerline and the outer limits of the specified width of the heated zone to a uniform pattern throughout the length of the joint.

Bibliography

1. Gourd, L.M., 1995, *Principles of Welding Technology*, Edward Arnold, London.
2. Dowling, P.J., Knowles, P.R., and Owens, G.W. (eds.), 1988, *Structural Steel Design*, The Steel Construction Institute, London.
3. Easterling, K., 1922, *Introduction to Physical Metallurgy of Welding*, Butterworth-Heinemann, Oxford.
4. Blodgett, O.W., 2002, *Design of Welded Structures*, The James F. Lincoln Arc Welding Foundation, Cleveland, OH.
5. Ghosh, U.K., 2006, *Design and Construction of Steel Bridges*, Taylor & Francis Group, London.

3

Defects in Welded Joints

ABSTRACT The defects attributed to internal sources such as the weld itself and the heat-affected zone (as against external sources such as fatigue effect and brittle fracture) are discussed. The defects in welds include undercut, porosity, slag inclusion, pin hole, incomplete root penetration, lack of fusion, solidification crack, and poor weld profile. The causes and the strategies for elimination of the defects are dealt with. Discontinuities in the load path and stress concentration are also discussed. A separate section has been devoted for typical defects in the heat-affected zone, namely, hydrogen or cold cracking and lamellar tearing along with measures to be adopted for their prevention.

3.1 Introduction

Ever since welding became popular as a method of connecting different components of a structure, defects in welded joints had been causing concern to designers and fabricators alike. Due to extensive research programs that were undertaken during the past few decades, there has been a greater understanding of these problems among scientists and engineers. Consequently, designers have now greater confidence in designing welded steel joints.

Defects in welded joints can be divided into the following two broad categories:

1. Those attributed to internal sources such as the weld and the heat-affected zone (HAZ)
2. Those occurring during service condition and attributable to external sources such as fatigue effect and brittle fracture

In this chapter, only the former category is discussed. The latter category, that is, brittle fracture and fatigue in welded joints, is discussed in Chapters 5 and 9, respectively.

3.2 Defects in Welds

Although it is assumed for the design purpose that welds will be perfect and free from any defect, this condition is seldom achieved in actual practice, and in spite of best efforts of all concerned, a few defects are likely to be present. These defects are mostly related to the choice of various welding parameters, incorrect profile of edge preparation, joint restraint, poor weld profile, rapid cooling rate, and, to a large extent, skill of the welder engaged in the work. Strict inspection of welded joints is, therefore, absolutely necessary as a part of the quality control regime. Quality control methods have been discussed in Chapter 6. Some typical defects that are commonly noticed in welded joints are discussed in the following paragraphs.

3.2.1 Undercut

A typical undercut in a welded joint is shown in Figure 3.1a. This defect is a potential danger in initiating fatigue crack, which is associated with

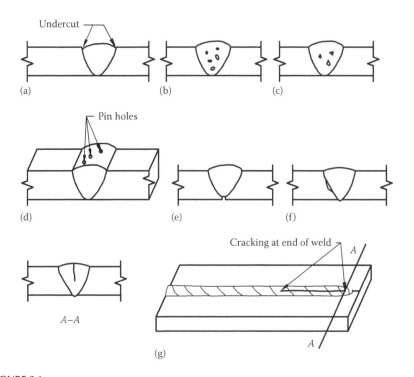

FIGURE 3.1
Typical weld defects: (a) undercut, (b) porosity, (c) slag inclusion, (d) pin holes, (e) incomplete root penetration, (f) lack of fusion, and (g) solidification cracking.

structures subjected to fluctuating loads, such as crane girders, offshore structures, and bridges (see Chapter 9). The possible causes of undercut in a joint are the following:

- Current too high
- Welding speed too high
- Arc length too long
- Incorrect manipulation of the structure being welded

The welding parameters need to be reviewed in order to eliminate this defect in a joint.

3.2.2 Porosity

Porosity is illustrated in Figure 3.1b. This defect, if present on the surface of the weld, may lead to tiny notches on the weld face and initiate fatigue cracks in the weld. The possible causes of porosity are the following:

- Wrong choice of electrode
- Welding speed too rapid
- Current too low
- Presence of sulfur or other impurities in the parent metal

The possibility of the defect may be reduced by using low-hydrogen electrode. Also, if the defect is noticed early, the welding parameters may be reviewed.

3.2.3 Slag Inclusion

Slag inclusion, as illustrated in Figure 3.1c, is initiated by flux left behind in the molten weld pool and may be particularly damaging when large inclusions or groups of small inclusions lie across the direction of the force, which might initiate fracture in the weld metal. The possible causes of this defect are the following:

- Weld temperature too low
- Cooling too rapid
- Included angle of the joint too narrow
- Viscosity of the molten metal too high
- Inadequate cleaning of slag between succeeding runs in multirun welds

In order to reduce the possibility of this defect, preheating of the components is recommended. Also, in multirun welds, care should be taken to clean the weld surface before succeeding runs are deposited.

3.2.4 Pin Holes

Pin holes are normally surface defects, as shown in Figure 3.1d. The probable causes are the following:

- Damp electrodes
- Presence of rust, scale, and paint on the surface of the parent metal

To eliminate this defect, dry electrodes should be used; if necessary, the electrode should be dried in ovens before use. Also, contaminants on the work need to be removed prior to commencement of welding.

3.2.5 Incomplete Root Penetration

Incomplete root penetration occurs when the weld does not reach the full depth of the edge preparation. The defect is illustrated in Figure 3.1e. The possible causes of this defect are the following:

- Root gap too small
- Root face too large
- Electrode size too large

To eliminate this defect, the profile of the edge preparation and also the welding parameters need re-examination.

3.2.6 Lack of Fusion

Lack of fusion occurs when the arc does not melt the parent metal prior to the weld metal touches it. As a result, the molten metal just rests against the parent metal instead of being integrated with it.

The defect is illustrated in Figure 3.1f. The possible causes are the following:

- Incorrect profile of edge preparation
- Arc not hot enough for the metal thickness
- Welding speed too rapid
- Current too low
- Electrode too large

To eliminate this defect, the profile of the edge preparation and the welding parameters need re-examination.

3.2.7 Solidification Cracks

Solidification cracks occur during the cooling process, typically at the center-line and at the end of weld deposit, as illustrated in Figure 3.1g. These cracks are often found in root runs and are normally visible after de-slagging. However, in some cases, these cracks may occur just below the surface of the weld and consequently not visible even after de-slagging. These cracks can be quite deep and seriously jeopardize the efficiency of a joint. The main causes of this defect are the following:

- Cooling too fast
- Wrong choice of electrode
- Incorrect profile of edge preparation
- Welds stressed during welding operation

The possibility of this defect can be substantially reduced by adopting the following strategies:

- Preheating as well as postheating of the parent components
- Using low-hydrogen electrodes
- Adopting correct profile of edge preparation
- Reviewing arrangement of jigs and fixtures

3.2.8 Defective Weld Profile

Defective weld profile includes erratically shaped, underfilled welds, over-laps, or excessive convexity. The primary causes are the following:

- Wrong welding parameters
- Lack of welder skill, practice, and diligence

Some of these defects and recommended acceptable weld profiles are shown in Figure 3.2. It may be noted that these defects are not due to the character-istics of the parent metal. Rather, the defects are related to the choice of the profile of the edge preparation, to welding parameters, and, to a large extent, to the level of the welders' skill. These aspects need to be kept in mind while trying to improve on the quality of the weld.

3.2.9 Issues Related to Defects in Welds

Primarily, the defects discussed in the foregoing paragraphs affect the behavior of the structure in two ways.

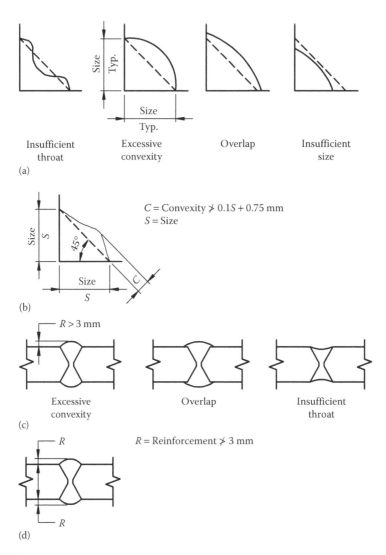

FIGURE 3.2
Defective weld profiles: (a) defective fillet weld profiles, (b) recommended acceptance criterion for fillet weld profile, (c) defective butt weld profiles, and (d) recommended acceptance criterion for butt weld profiles.

3.2.9.1 Discontinuity in the Load Path

The direct effect of the defects is discontinuity in the load path and reduction in the load-carrying area of the joint, leading to impairment of the strength of the joint. However, thankfully, except for incomplete penetration, the other defects mostly occur at isolated locations and are seldom concurrent at a particular cross section. Therefore, chances of significant loss of area along the entire cross section are rather remote. Nevertheless, acceptability of such defects depends on the size and nature of the defects in a particular location and has to be reviewed for each individual case. With regard to incomplete penetration, the defect is likely to occur along a considerable portion of the weld length. Therefore, the defect warrants careful and close examination. Figure 3.3 illustrates stress paths and locations of typical failures in plates joined by butt welds.

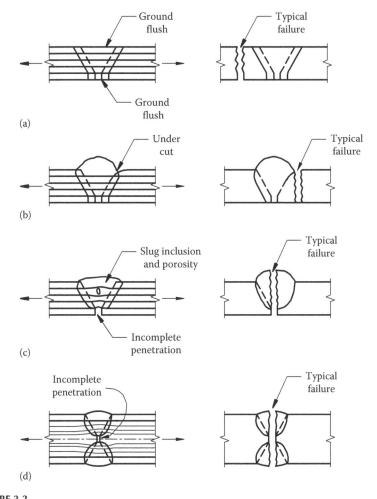

FIGURE 3.3
Stress paths and locations of typical failures in plates joined by butt welds: (a) no defect; (b) undercut; (c) slug inclusion, porosity, and incomplete penetration; and (d) incomplete penetration.

3.2.9.2 Stress Concentration

Stress concentration at the location of discontinuities of load path is of par-ticular concern to the designer. However, chances of stress concentration for smooth and rounded discontinuities such as porosity and slag inclusion are considerably less than those for sharp notches such as undercuts or even tiny cracks. Stress concentration has a significant impact on the fatigue strength of a joint. The topic is discussed in Chapter 9.

3.3 Defects in HAZ

Some of the typical defects occurring in HAZ that are of particular concern to structural engineers are discussed in the following paragraphs.

3.3.1 Hydrogen Cracking or Cold Cracking

Arc welding operation invariably involves building up a thermal gradient in the parent metal. The severe thermal cycle—heating and then rapidly cooling—in the HAZ changes the metallurgical structure of steel in the region from a ductile to a hard form, making it prone to cracks due to residual stress present in the weld metal. To add to this, a substantial portion of hydrogen present in the molten weld pool diffuses into the HAZ and induces formation of cracks in the hardened microstructure, which is already susceptible to cracking. This type of defect is often termed *cold cracking* because it occurs generally when the metal has cooled to ambient temperature. Hydrogen cracks may also occur in the weld metal, but this defect is not common.

In order to prevent hydrogen cracking, the following measures are recommended:

- Preheating of the components and welding heat input should be bal-anced so as to reduce the rate of cooling and thereby avoid quench-ing the HAZ to high hardness.
- Low-hydrogen electrodes and/or other consumables should be used to minimize the amount of hydrogen in the weld pool.
- Metal surfaces should be cleaned properly in order to remove grease, paint, moisture, and so on.
- High restraint weld geometry should be avoided in the detailing of the joint to minimize residual stress.

3.3.2 Lamellar Tearing

Theoretically, structural steel is considered to be homogeneous material; that is, the steel material is uniform in its properties over its volume. On this basis, steel is assumed to be an isotropic material, that is, a material possessing the same elastic strength and properties in all the axes. In actual practice, however, during the rolling process, minute islands of sulfides and other compounds (in the form of flattened out inclusions) are produced in some steels that result in platelets (laminae), typically running parallel to the surface. As a result, the plane running parallel to the direction of the rolling is likely to lack cohesive strength. Thus, the load-carrying capacity in the transverse direction (i.e., through-thickness direction) at some locations is significantly reduced. This defect in the steel material is termed *lamination*. It can be detected by nondestructive testing such as ultrasonic testing.

In welded joints, as the weld metal cools, it contracts producing high stress in the transverse direction. When this resultant stress is carried in the through-thickness direction, the presence of lamination may cause the plate to be separated. This phenomenon is termed *lamellar tearing*. The separation has a step-like appearance, comprising of a series of terraces parallel to the surface (horizontal and vertical cracking) of the base metal. Lamellar tearing is generally associated with the welding of highly restrained structures, such as rigid box girders, tee joints, and corner joints. A typical lamellar tearing in cruciform welded joints is illustrated in Figure 3.4.

Summarizing, the factors that are most likely to cause lamellar tearing are the following:

- Poor through-thickness properties of the parent metal (i.e., low load-carrying capacity in the transverse direction) due to the presence of long, sharp nonmetallic inclusions lying in layers parallel to the surface
- Inappropriate design of the weld joint configuration, which induces high residual tensile stress in the transverse direction of the parent metal
- Use of thick plates

In order to reduce the chance of lamellar tearing in welded joints, particularly in cruciform, tee, or corner joints, it is recommended to use steels with good

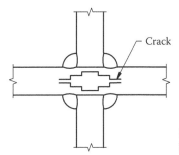

FIGURE 3.4
Lamellar tearing.

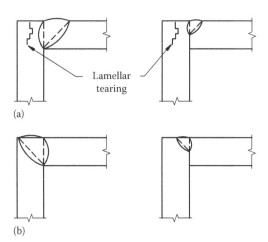

FIGURE 3.5
Susceptible (a) and improved (b) details to avoid lamellar tearing.

through-thickness properties. Also, the related plates should be tested by ultrasonic test prior to commencement of fabrication process, and suspect plates with risk of lamellar tearing be replaced by plates with adequate through-thickness properties. Also advisable is to design and detail the joints in thick plates avoiding weld contraction in the direction normal to the surface of the plates, that is, through-thickness direction. Furthermore, heavy fillet welds should be avoided in the design. Figure 3.5a shows examples of details of corner joints that are susceptible to lamellar tearing. Corresponding improved details are shown in Figure 3.5b, in which the modified edge preparation detail reduces the risk of lamellar tearing.

3.4 Concluding Remarks

Defects in welded joints are almost an inseparable part of the welding culture. However, these defects can be reduced by a number of methods. One of them is slower cooling rate. This can be ensured by preheating as well as postheating the joining components. Use of appropriate parameters (current, welding speed, etc.) compatible with the profile of the edge preparation, selection of correct electrode (e.g., low-hydrogen electrode), and detailing of correct profiles of edge preparation are the other essential features of good-quality welded joints. Of course, the necessity of the welders' skill cannot be overemphasized. To reduce the stresses in the weld during welding operation, it may also be necessary to review the design and arrangement of jigs and fixtures. Quality control plan and associated inspection system for welded joints are discussed separately in Chapter 6.

Bibliography

1. Hicks, J., 2001, *Welded Design—Theory and Practice*, Abington Publishing, England.
2. Easterling, K., 1992, *Introduction to Physical Metallurgy of Welding*, Butterworth-Heinemann, London.
3. Houldcroft, P., and John, R., 1988, *Welding and Cutting*, Woodhead-Faulkner, England.
4. Gourd, L.M., 1995, *Principles of Welding Technology*, Edward Arnold, London.
5. Xanthakos, P.P., 1994, *Theory and Design of Bridges*, John Wiley & Sons, New York.
6. Blodget, O.W., 2002, *Design of Welded Structures*, The James F. Lincoln Arc Welding Foundation, Cleveland, OH.
7. Evans, J.E., and Iles, D.C., 1998, *Guidance Notes on Best Practice in Steel Bridge Construction*, The Steel Construction Institute, Ascot.
8. *Code of Practice for Metal Arc Welding in Structural Steel Bridges (Welded Bridge Code)*, 2001, Research, Design and Standards Organisation, Indian Railways, Lucknow, India.

4

Control of Welding Distortion

ABSTRACT The chapter begins with the causes of distortion and its types in welded structures, followed by discussion on preventive measures, in both design and fabrication stages. Practical guidelines on various methods for correcting different types of distortion after fabrication have also been deliberated.

4.1 Introduction

Arc welding method involves a moving heat source, which forms a molten weld pool along the joint leading to a changing pattern of temperature (heating-cooling cycle) and strain. As the metal experiences progressive heating and cooling, heat flows from the pool into the adjoining parent metal, thereby causing:

- Change in the properties of the metal
- Distortion to the components being joined

The former, that is, the change in the properties of the metal, has been discussed in Chapter 2. In this chapter, characteristics of the distortion due to thermal expansion and contraction during welding and its control are discussed. The suggested methods for control of distortion are based on practical experience in the shop floor with adequate analytical support.

4.2 Basic Causes of Distortion

The basic causes of distortion in steelwork may be attributed to many factors. The most important factors are discussed in the following paragraphs.

4.2.1 Properties of Materials

The coefficient of thermal expansion of the parent material and weld metal has significant influence on distortion. In a welded joint, as the temperature rises, the coefficient of thermal expansion also increases. The higher the coefficient of expansion, the greater would be the tendency to distort. Also, distortion in stainless steel fabrication would be more than that in an equivalent fabrication in mild steel. This happens because of the higher coefficient of expansion of the former.

The other aspect that influences distortion is that the yield point of steel is often reduced drastically as the temperature rises to the range of 500–600°C.

4.2.2 Inherent Stresses in Parent Material

It is quite common that some inherent stresses are present in the components before these components are welded. These stresses generally originate during rolling operation in the steel plant and during prewelding fabrication stage (e.g., shearing, bending, etc.). The heat applied to these components during subsequent welding process tends to relieve these stresses, which may increase or reduce the distortion due to welding. Thus, these stresses may play a significant part in the mechanics of distortion.

4.2.3 Uneven Heating

Steel material expands in all directions when heat is applied *uniformly* and the material is *not restrained*. Similarly, it will contract and return to its original form if allowed to cool uniformly without any restraint. However, if the applied heat is concentrated in one area only, the heating will not be uniform and the expansion would be local and therefore uneven. As a result, on cooling, uneven contraction will occur, causing the components to distort. The magnitude of distortion would depend on the heat input, the nature of heat source, and the way the heat is applied.

Heat, other than that applied by welding process, may also influence distortion. Thus, preheating, applied wrongly, may increase distortion.

4.2.4 Restraint during Welding

If no external restraint is applied to the components during welding process, and subsequent cooling, the fabricated structure is likely to distort because of internal stresses developed due to welding. On the other hand, if restraint is imposed on the components and distortion is not allowed to take place, the weld and the parent metal become stressed during welding and cooling.

In the case of arc welding, the problem becomes rather complex because of the presence of an *intense* concentrated heat source, which is also *moving*. As the heat source moves forward along the joint axis, the weld pool metal at the trailing end cools. This changing pattern of temperature subjects the metal to experience progressive intense heating leading to melting, then rapid cooling, and simultaneous solidification. The cooling and solidification starts from the fusion boundary and progresses inward in the direction

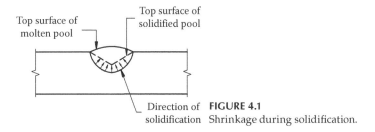

Top surface of
molten pool

Top surface of
solidified pool

Direction of **FIGURE 4.1**
solidification Shrinkage during solidification.

of the centerline (Figure 4.1). As the density of the solidified metal is higher than that of the liquid metal, the solidified metal at the lower portion occupies lesser volume than the liquid metal at the upper portion, that is, it shrinks. This lower solidified portion also shrinks due to thermal effect. Simultaneously, the liquid metal at upper portion also shrinks due to thermal effect during cooling process. As the weld metal shrinks, due to the combined effect of density and thermal contraction (restraint), the joint as a whole tends to contract. The surrounding cooler material tries to prevent the contraction restraint, thereby inducing locked-up stresses (residual stress) in the weld and producing plastic deformation. If yield point of the metal is reached during this deformation, the distortion will be of permanent nature and the structure will not come back to its original shape even if allowed to cool.

4.3 Types of Distortion

The weld metal experiences shrinkage both in the direction of the weld (longitudinal) and at right angles to it (transverse). These two types of shrinkages cause the components being welded to suffer distortion, which may be classified into following types:

- Longitudinal distortion
- Transverse distortion
- Angular distortion
- Bowing (mainly in thin plates)

The different types of distortion are illustrated in Figure 4.2.
Longitudinal shrinkage of weld reduces the length of the joining plates at the line of the welded joints, thereby causing longitudinal distortion. Similarly, due to transverse shrinkage of the weld, the overall width of the plate is reduced and consequently transverse distortion takes place.

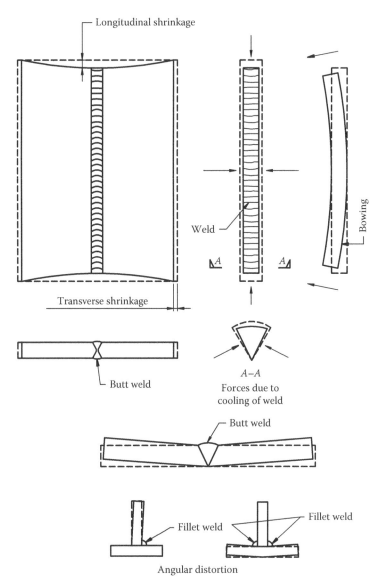

FIGURE 4.2
Types of distortion.

Furthermore, in case of an asymmetrical butt welding, such as in a single-V butt weld, the transverse shrinkage at the top surface of the weld will be greater than that at the root level. This differential shrinkage is due to the fact that contraction in weld is proportional to the volume of the metal being cooled. In this case, the volume of weld metal in the transverse direction at the top level is more than that at the bottom level. Thus, the transverse shrinkage at the top level is more than that at the bottom level. This will produce a rotational or angular distortion.

In the case of welding thin plates, simultaneous longitudinal and transverse contraction often leads to bowing, as illustrated in Figure 4.2.

4.4 Control of Distortion

There are two options to ensure that the finished welded structure is free from distortion:

1. Prevention
   The structure is so designed and so welded that its final shape and dimensions are within the tolerance limits specified in the guiding standards.
2. Correction
   The structure is allowed to distort during welding followed by post-fabrication correction regime to bring the final structure within the tolerance limits.

Either of these methods or a combination of both may be used depending on various factors. As is expected, each structure will have its individual features requiring adoption options to be considered. However, prior knowledge about the factors that affect distortion is very important for deciding on the method to be adopted for keeping distortion within permissible limits.

4.4.1 Prevention of Distortion

Awareness of the basic causes of distortion is essential for all persons and agencies connected in the welded steel structures, namely, the designer, the detailer, the welder/operator, and the supervisor, so that each in their area may contribute to prevent or reduce the chances of distortion. In this section, the precautions that should be taken from the design stage to the completion of fabrication are briefly discussed.

4.4.1.1 Design Stage

In the initial stage of design, it should be considered to incorporate rolled sections as much as possible in preference to built-up sections. This option will not only be economical but also reduce the chances of distortion. Consideration should also be given to use the longest and the widest plates within the working capacity of the shop, so as to use minimum number of welded joints in fabrication.

To reduce chances of distortion, it is necessary to reduce the effective shrinkage force. To achieve this, the weld size must be kept to the minimum to meet the service requirement. For example, the effective throat thickness determines the strength of a fillet weld. Thus, depending on the requirement or importance of the joint, partial penetration welds may be considered in preference to full penetration welds. Also, making the weld profile too convex does not increase the strength but certainly increases the effective shrinkage force. A flat or concave profile with adequate throat thickness will reduce the shrinkage force without impairing the strength.

Similarly, in the case of butt weld, the adoption of a proper edge preparation will reduce the effective shrinkage force. The aim should be to obtain the proper fusion at the root of the weld with minimum weld metal. A bevel up to 30° should satisfy the requirement. Also, the gap between the two pieces should be kept to the specified minimum, so that the least amount of weld metal is deposited and consequently the shrinkage force is kept to the minimum.

Distortion in the lateral direction can be kept to the minimum by using less passes with the largest suitable size of electrode, bearing in mind the thickness of the pieces, welding position, length of run, and so on.

Welded joints should be located in such a manner that these are easily accessible for welding. Particularly, these should be fully visible to welders to enable them to deposit proper weld *without excessive heat input*.

It must also be borne in mind that the chances of distortion increases with the decrease of plate thickness and becomes more pronounced in light structures.

1. Butt welded joints
   a. Transverse and longitudinal shrinkage and bowing
      In a butt welded joint involving multiple runs of weldment, the first root run tends to pull the plates together, and because of the root gap, they are free to move. The second run, in turn, tries to do the same but is restrained by the first run beneath it, which is compressed by the upper run. This pull at the top and push at the bottom of the weld tends to produce an angular distortion. The initial contraction of the first run and the tendency of the subsequent runs to contract together result in a transverse shrinkage in the joint. Thus, the total shrinkage is a function of the number of runs involved in a butt weld rather than the thickness of the components joined. Therefore, in order to reduce transverse shrinkage, the largest practical gauge of electrodes should be used, thereby minimizing the number of runs.

Forces similar to those discussed in the previous paragraph exist along the length of the joints also. These forces will produce longitudinal shrinkage and in some cases bowing. However, longitudinal shrinkage is much lower in magnitude compared to the transverse shrinkage.

b. Angular distortion

As contraction on cooling is proportional to the length of the metal being cooled, in asymmetrical welds such as single-V butt welds, shrinkage in the transverse direction will be greater at the top surface than at the bottom (root). This will produce a rotational (angular) distortion. This is illustrated in Figure 4.3.

There are a number of methods for keeping the angular distortion within an acceptable limit. One of them is to clamp the pieces to restrain the movement. This method is likely to produce *locked-in* stresses in the joint. An alternative method is to balance the welding about the neutral axis, that is, to adopt a weld preparation in which the shrinkage of the plates across the thickness of the weld would be more or less the same. Generally, a double-V or double-U joint may provide such a solution (Figure 4.4). However, in a multirun double-V butt weld, the first run is likely to produce more angular rotation than the subsequent runs. Thus, symmetrical double-V preparation may not produce a completely distortion-free result. A slightly asymmetrical preparation (Figure 4.5) is likely to produce satisfactory results.

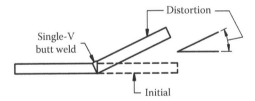

FIGURE 4.3
Rotational (angular) distortion.

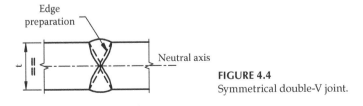

FIGURE 4.4
Symmetrical double-V joint.

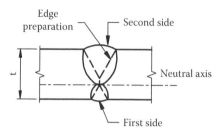

FIGURE 4.5
Asymmetrical double-V joint.

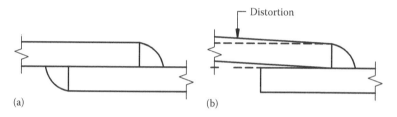

FIGURE 4.6
Lap joints: (a) double fillet weld and (b) single fillet weld.

>   In this case, welding in the minor groove is completed before
>   starting the major groove.

2. Fillet welded joints

>   In order to reduce distortion in fillet welded joints, it is imperative
>   that while having regard to the weld strength, the weld size (and
>   thus the heat input) is kept to the minimum.
>
>   Lap joints are frequently used in large arc-welded plate struc-
>   tures. This type of joint does not present major distortion-related
>   problems, particularly if double fillet welded joints, as shown in
>   Figure 4.6a, is used. These joints are generally liable to longitudi-
>   nal shrinkage, which, in case of long seams, may cause buckling. In
>   these joints, transverse shrinkage or angular distortion is negligible.
>   Single-fillet lap joints, as shown in Figure 4.6b, are susceptible to dis-
>   tortion and should be avoided.

4.4.1.2 Fabrication Stage

It is imperative that the component parts are accurately fabricated to the
required size and form, to ensure that the welded product will have the
shape and dimensions specified in the fabrication drawings. Initial inaccu-
racies in size and form are likely to entail poor fit-up of the components

(e.g., joint gap may vary along the length of the seam). In a butt welded joint, too large a gap will require more weld metal to make the joint, whereas too small a gap may cause lack of penetration. Also, if the components are forced into position by mechanical means, internal stresses may be developed that might add to the stresses set up by welding in the weld and the adjoining material leading to increased distortion. Thus, accuracy in fabrication of the components is of utmost importance for the prevention of distortion during fabrication.

For countering distortions and bowing during fabrication stage, there are a few common methods that are discussed in the following paragraphs:

1. Presetting method

    In this method, the components to be welded are preset to compensate for the distortion. An example of presetting method is illustrated in Figure 4.7. In the joint shown in Figure 4.7a, the components are assembled in the exact relationship as will be required after welding, resulting in angular distortion. If the components are preset at predetermined angles as shown in Figure 4.7b, and then welded, the shrinkage in the weld during cooling will bring the components to their correct relative positions and the joint will be free from angular distortion. Needless to add the degree of presetting requires sufficient field experience and practice to be successful. In case of butt welded joints in plates having free

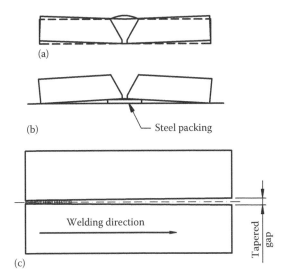

FIGURE 4.7
Presetting of butt joints: (a) not preset, (b) preset with steel packing and (c) preset with tapered gap.

edges, the shrinkage of the weld tends to pull the joint faces closer as the weld progresses. To avoid this trouble, the gap between the weld faces of the plates may be preset in a taper, as shown in Figure 4.7c.

In composite beams where shear connectors are attached in the top flange for composite action with the concrete slab, cover plates are often fillet welded to the bottom flange to achieve increased strength. Figure 4.8 illustrates a typical layout. As the welds cool, they tend to shrink and cause the beam to distort (bow) in the plane parallel to the web. If the width of the cover plate is less than the width of the flange plate, the welding is normally done by down-hand welding with the beam supported near its midpoint and its ends cantilevered out (see Figure 4.8a). This arrangement will reduce the distortion. If the cover plate is wider than the width of the flange, downhand welding can be done in the normal position, and in this case, the beam should be supported at the ends to reduce the distortion. Figure 4.8b illustrates the arrangement.

Angular distortions have often been noticed in flanges of fillet welded plate girders (Figure 4.9). This type of distortion can be controlled by prebending the flange plate prior to welding. An alternative solution to this problem is to deposit the fillet welds in a

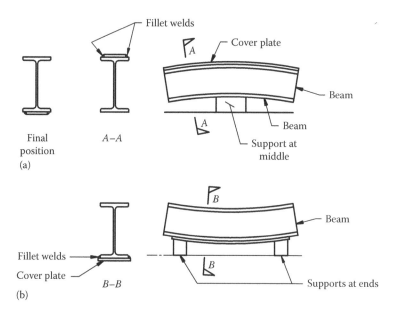

FIGURE 4.8
Presetting of beams with cover plate welded on one flange only: arrangement for welding (a) when width of cover plate is less than bottom flange width and (b) when width of cover plate is more than bottom flange width.

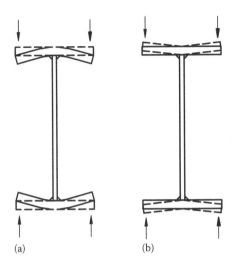

FIGURE 4.9
Prebending.

predetermined sequence. As the angular distortion is maximum in the first weld (on one side of the joint), the second weld (on the other side) cannot pull back the plate fully to the normal position. Thus, a residual distortion still remains in the joint. To minimize this defect, the joint can initially be set at a predetermined angle so that at the end of the second weld, the required angle of 90° is achieved. The procedure is illustrated in Figure 4.10. In plate girders where automatic welding process is employed, stiffeners are not commonly put in first, because they are likely to impede the progress of the automatic machine. Where the flange plates are relatively thin, angular distortion of these plates may occur. In such cases, it may be considered necessary to prebend as well as preset the flange plates to ensure flatness of the flange plates after welding. As in other methods of presetting, this method also requires considerable experience to be successful.

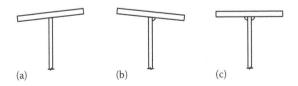

FIGURE 4.10
Presetting: (a) before welding, (b) after first weld, and (c) after second weld.

2. Restrained assembly method
   In this method, suitable clamps, jigs, or fixtures are used to control
   distortion during welding operation within tolerance limits. Often
   component parts are suitably tack welded before welding, thereby
   applying restraint to prevent distortion.

   Restrained assembly method is likely to produce *locked-in* stresses
   (residual stresses) at the joint. In most cases, these stresses are not
   significant. However, in cases where these stresses are likely to be
   high, suitable welding sequence and/or preheating the components
   may solve the problem. Where service conditions preclude the presence of residual stress, a stress-relieving heat treatment should be
   applied after welding.

   It is not desirable that movement of the components is completely
   restrained during welding. Overall movement should be controlled
   in a balanced manner, such as to restrain movement in one direction while allowing freedom in another. Figure 4.11 illustrates a
   restraining arrangement known as *strongback* to prevent angular
   distortion in a butt weld joint, while allowing it to shrink freely in

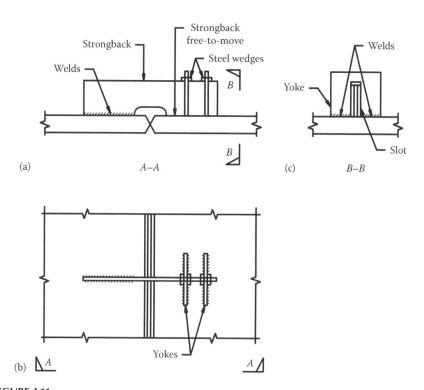

FIGURE 4.11
(a–c) Typical detail of strongback attachment in a butt weld joint.

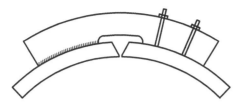

FIGURE 4.12
Typical detail of strongback attachment in a butt weld joint of cylindrical shell.

the transverse direction. It should be ensured that the strongback arrangement should never be welded rigidly on both sides of the joints. Figure 4.12 illustrates a strongback arrangement for single-V butt weld joint in a cylindrical shell. This is a common method for site welding of butt joints in storage tanks and pressure vessels. Strongback and wedge system can also be used to compensate for angular distortion in single-V butt weld joints of thin plates, as illustrated in Figure 4.13.

3. Higher welding speeds
   Higher welding speeds by using semiautomatic or fully automatic arc welding equipment generally reduce distortion.

4. Welding sequence
   Distortion can be controlled during welding by balancing the shrinkage forces with other forces. This can be achieved by employing proper welding sequence, so that shrinkage of weld metal at one location will counteract the shrinkage forces of welds already made at another location. A simple example of this situation is welding alternately on both sides of the neutral axis of a double-V preparation in a butt welded joint. Also, for a long joint, the welding should commence from the center and proceed toward the free ends. Figure 4.14 illustrates a method, known as *step-back* method, which will further minimize distortion.

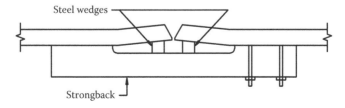

Steel wedges

Strongback

FIGURE 4.13
Typical detail of strongback with steel wedges in a butt weld joint of thin plates.

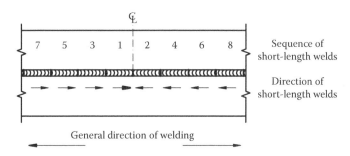

FIGURE 4.14
Typical sequences and directions of welds in butt weld joints to reduce longitudinal distortion.

In any complicated type of work, it is necessary to prepare a schedule showing the order in which welds are to be deposited along with the direction of welding. Such a schedule may have to be modified as the fabrication work progresses and experience is gathered.

5. Shrinkage allowance
   As shown in Figure 4.2, shrinkages both in longitudinal and in transverse direction take place in all welded fabrication work. Consequently, it is imperative that shrinkage allowances for these welded fabrication work should be made at the assembly stage. In effect, the components (to be welded) are to be fabricated oversize by an amount sufficient to cover the shrinkage. Materials are to be procured accordingly. Furthermore, the welding fixtures must also be designed considering this aspect.

4.4.2 Correction after Fabrication

In practice, in spite of all efforts, controlling of distortion within permissible limits is not always possible. In fact in certain cases, it may be more practical and also economical to allow the structure to distort during welding and to correct it afterward. In most cases, it is possible to rectify the distortion, provided the product has not shrunk so much that it cannot be used even after allowing the dimensional tolerances.

In general, there are two methods being used to rectify distortion:

- Mechanical means
- Heating

4.4.2.1 Mechanical Means

In case a reasonably small light structure has bowed after welding, it can be straightened on a hydraulic press by inserting suitable packers between the

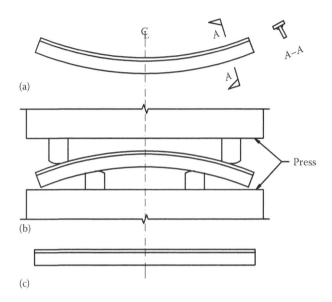

FIGURE 4.15
Correction of bowing by use of press: (a) distortion after welding, (b) distorted member in press, and (c) member after pressing.

structure and the press. This is illustrated in Figure 4.15. Another method is to pass the light structure through *straightener* machine between their jaws to straighten it.

A distorted flat fabricated member can often be corrected by clamping it to some strong structure with flat surface and then stress relieving the distorted structure by heating slowly to 650°C and maintaining the temperature for about 1 hour for every 25 mm of thickness. It should then be allowed to cool by itself without any external help.

Peening of individual weld run in heavy butt joints immediately after deposition stretches the weld bead, thereby counteracting the tendency to contract and shrink as it cools. However, peening should be done carefully, so as not to damage the weld metal. Final layers should never be peened.

4.4.2.2 Correction by Heating

Concentrated heat from a blow pipe can be used for straightening flange plates distorted during welding by applying the simple concept of thermal expansion and contraction. Although the method appears to be quite simple, it needs sufficient field experience and practice to be successful. As a result, though very effective, the method has not gained popularity among technicians. Figure 4.16 illustrates this principle. When a narrow band on top face of top flange is heated by applying concentrated heat from a blow pipe, the

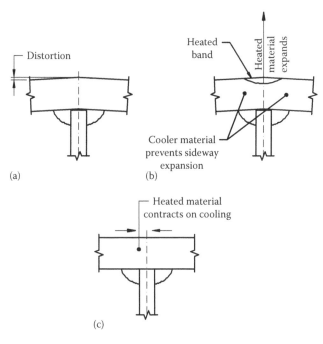

FIGURE 4.16
Flame straightening: (a) distortion in flange, (b) heating along joint line, and (c) flange straightened.

heated portion will tend to expand sideways. This expansion will be prevented by the surrounding cooler material, and the relatively weak heated part is forced to expand at right angles to the surface. When the area cools, the metal contracts. This contraction causes the band to straighten in a manner similar to which caused the distortion due to weld. With regard to temperature, it should be restricted to approximately 700°C, that is, a dull red. Simultaneously with heating on the top, cooling by means of water jet may be applied on the reverse side. When a thin plate is welded to a considerably rigid flange, the plate is likely to distort in the manner shown in Figure 4.17. Such distortion may be removed by heating the plate in local spots on the convex side. Spot heating should be done systematically starting at the central region of the distortion and proceeding outward toward the rigid frame.

In cases of heat application other than spot heating of a thin plate, it is general practice to apply heat in a wedge-shaped manner, as shown in Figure 4.18. Heating should proceed from the base to the apex of the wedge. It should penetrate evenly through the plate thickness, maintaining a uniform temperature. Applying the wedge-shaped heating pattern may be used to rectify a fabricated beam, which has been cambered due to welding. Similarly, a cambered plate may be straightened by applying the same principle.

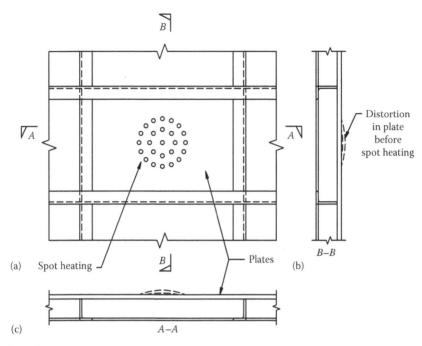

FIGURE 4.17
(a–c) Typical pattern of spot heating to correct distortion in plate.

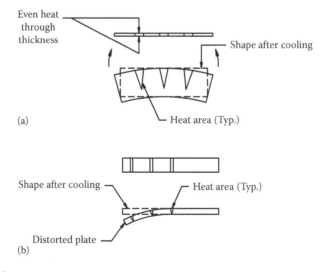

FIGURE 4.18
Local heating to straighten camber and distortion of plates: (a) plate with camber and (b) plate with distortion.

4.5 Concluding Remarks

Fabricated structures are designed to perform specified duties and must conform to the specified dimensions and tolerances. It is imperative therefore that distortions are foreseen and controlled or corrected; otherwise, the product may not be able to serve its intended purpose. Discussions in this chapter deal with the general principles that may be applied in the various stages of design and manufacture and offer practical guidelines for those involved in welded steel structures.

Bibliography

1. Gourd, L.M., 1995, *Principles of Welding Technology*, Edward Arnold, London.
2. Easterling, K., 1992, *Introduction to Physical Metallurgy of Welding*, Butterworth-Heinemann, Oxford.
3. Houldcroft, P., and John, R., 1988, *Welding and Cutting*, Woodhead-Faulkner, Cambridge.
4. *Control of Welding Distortion*, 1957, The Institute of Welding, London.
5. Allen, J.S., 1958, *The Use of Arc Welding in Structural Engineering*, in the Structural Engineer, October, 1958, Institution of Structural Engineers, London.
6. Blodget, O.W., 2002, *Design of Welded Structures*, The James F. Lincoln Arc Welding Foundation, Cleveland, OH.
7. Ghosh, U.K., 2006, *Design and Construction of Steel Bridges*, Taylor & Francis Group, London.
8. Hicks, J., 2001, *Welded Design—Theory and Practice*, Abington Publishing, England.
9. Samanta, A.K., 2009, *Guidebook for Fabrication and Erection of Steel Structures*, Institute for Steel Development and Growth, Kolkata, India.

5

Brittle Fracture

ABSTRACT The characteristics of brittle fracture in welded joints are discussed, along with the factors that influence it, namely, metallurgical feature, temperature of steel, and service conditions. The major areas that need special attention in welded structures to minimize the risk of brittle fracture are also discussed. The chapter ends with a case study of failure of welded steel girders due to brittle fracture.

5.1 Introduction

Brittle fracture in steel is characterized by failure of the material as a result of rapid crack propagation with very limited plastic deformation at a stress level below the yield stress of the material. Internal stresses due to thermal contraction enhance the risk of brittle fracture. As the energy required for initiating brittle fracture is rather low, such crack can be initiated from internal residual stress from welding without even any externally applied force.

The metallurgical mechanisms of brittleness in steel are complex and do not come under the purview of this text. However, for the purpose of the structural designer, it would be useful to understand that brittleness is characterized by a fracture that propagates rapidly at a relatively low tensile stress, and at low temperature, and is often associated with little or no prior plastic deformation. Propagation of such crack requires much less energy compared to a ductile crack, which is preceded by a considerable plastic deformation until the load exceeds the value corresponding to its yield stress. Thus, brittle fracture may occur at an applied load much lower than at which failure would normally be expected and usually starts from areas of high local stress concentration, such as welding crack and notch.

5.2 Factors Influencing Brittle Fracture

The factors that generally influence brittleness in steel are discussed in the following paragraphs.

5.2.1 Metallurgical Feature

Depending on their chemical composition, heat treatment, or mechanical working during production, some steel are more brittle than others.

5.2.2 Temperature of Steel in Service

Structural steel undergoes a ductile-to-brittle transition as the temperature falls, causing a change in their fracture behavior with temperature. This change in fracture behavior is largely influenced by the chemical composition of the steel used and its metallurgical structure. Thus, fracture may occur at even low stresses when the ambient temperature drops, say below the freezing point. Geographical location of the structure is therefore important from this point of view.

5.2.3 Service Conditions

Certain distribution pattern of force field tends to make steel susceptible to brittle fracture. Examples are locations of stress concentrations due to abrupt change in section, notches, and cracks. Also, thick and wide plates and deep webs in plate girders are subjected to higher degree of restraint causing complex internal stress pattern when loaded. This exposes them to higher risk of brittle fracture. The other causes of brittleness are cold working on steel during fabrication and also rapid rate of loading (impact).

5.3 Prevention of Brittle Fracture

The risk of brittle fracture in a welded structure can be effectively reduced by taking the following steps.

5.3.1 Selection of Appropriate Steel Material

For welded structures, the parent material should have adequate strength and notch toughness properties; that is, it should be adequately ductile at the service temperature.

Fracture mechanics test, Crack tip opening displacement (CTOD) test, and Charpy V-notch test are the most common. The first two are large-scale laboratory tests on samples of full thickness, while the third, that is, Charpy V-notch test, is a small-scale laboratory test on a small machined sample. This test is very popular for its simplicity, quick results, and reasonable cost. In this test, the energy absorbed by fracturing a specimen is the measure of fracture toughness of the steel. A specimen of a small square bar with a machined notch across the center of one side is hit

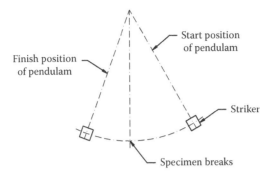

FIGURE 5.1
Schematic arrangement showing Charpy V-notch test.

on the face behind the notch by a striker mounted at the end of a swinging calibrated pendulum (Figure 5.1). This is done by first lifting the striker to a specified height, and then releasing it to hit and break the specimen, rising to a height on the other side. The energy absorbed in breaking the specimen is indicated over a scale by a pointer moved by the pendulum. This value is termed *impact value* and measures the fracture toughness of the steel. The test is carried out on a number of samples at different temperatures, such as room temperature, −10, −20, −40, +10, +20, and +40°C. The energy absorbed is found to vary with temperature. The range of temperature over which the change of energy occurs is termed *transition temperature range*. Steels of different levels of fracture toughness and different transition temperature ranges can be produced by the steel mills by adopting different combinations of chemical composition, heat treatment, and mechanical working. The lowest temperature at which the fabricated structure can be used is generally guided by the properties of the parent material, provided that the properties of the welded joint and the heat-affected zone are also, at least, up to the same level. In spite of its positive features, the Charpy test has its limitations as well.

Its results cannot always be interpreted in a quantitative way. However, its simplicity, quick results, and comparatively low cost make it a popular and valuable tool for selection of appropriate steel material.

From the foregoing, it is apparent that understanding the change behavior of the fracture toughness of steel (ductile to brittle stage), vis-à-vis transition temperature range, is an important aspect to be noted by the designer for selecting the appropriate material for the structure, which should be ductile at the required operating temperature during service condition.

5.3.2 Design of Details

Design of details plays a very important role in reducing the risk of brittle fracture in any steel structure, particularly in a welded structure. For example, fillet welds across the tension flanges of girders should be avoided as these fillet welds induce brittleness in the weld region. Intermittent welding should be avoided during detailing stage. Every time the arc strikes the parent metal, it runs the risk of being brittle. Also, as discussed earlier, brittle fracture is initiated almost always at points of stress concentration. Therefore, notches and sudden geometric changes, such as re-entrant angles, must be eliminated in any detail and large radii should be used at changes of sections. The details of the structure and its welded joints should be so developed as to ensure, as far as possible, smooth flow of stress pattern.

5.3.3 Quality Control during Fabrication

In order to reduce the risk of brittle fracture, certain aspects of fabrication work need to be carefully noted. Weld defects, such as undercutting, slag inclusion, porosity, or crack, enhance the risk of brittle fracture. It is therefore necessary to carry out adequate inspection regime during fabrication work, supported by nondestructive testing. Care should also be taken to avoid small fillet weld on relatively heavy members, as these then become susceptible to cracks. Furthermore, it should be noted that preheating of the components to be welded reduces the chance of brittle fracture. It is, therefore, necessary to develop an appropriate welding procedure including sequence of welding operations for reducing the chances of brittle fracture.

5.4 Learning from Failures

Instances of failure of steel structures due to brittle fracture have been demonstrated over the years in the form of some unfortunate accidents. Some examples of casualties of brittle fracture in welded steelwork are *Liberty Ships* during World War II; King's Bridge across River Yarra, Melbourne, Australia; and the Sea Gem drilling rig for North Sea gas. One such case study is discussed next.

One of the most well-known examples of brittle fracture failure in steel girders is the collapse of King's Bridge across River Yarra, in 1962 due to failures of welded tension flanges of 30 m span simply supporting welded plate girders. Each plate girder was built up by welding two flange plates and a web plate. An additional flange plate was welded to the underside of the bottom flange, in the central part of the span to cater for increased bending moment at this location. Figure 5.2 illustrates the typical details of the plate girders. On July 10, 1962, barely after 15 months of use, four girders failed due to fracture after the passage of a 45-ton vehicle, although it was well within the permissible limit. It was revealed during investigation that the cracks in most cases started in the bottom flanges close to the transverse welds at the

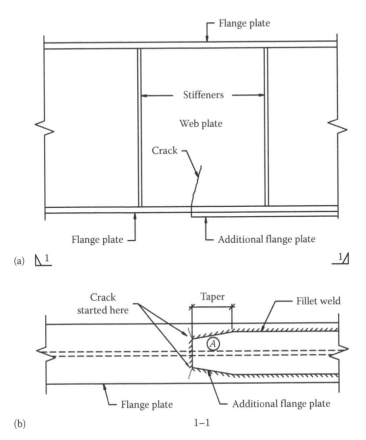

FIGURE 5.2
(a, b) King's Bridge, Melbourne, Australia.

ends of the additional cover plates. The cracks progressed upward through the flange plate toward the web plate—in some cases, severing the entire girders. Investigations also revealed that the transverse welds at the ends of the additional cover plates (location A in Figure 5.2) were welded last, after the completion of the longitudinal welds, causing complete restraint against contraction when the transverse weld was deposited, and thereby resulting in high residual stress. This made the area particularly prone to transverse cracking. Apparently, this aspect was overlooked during detailing as well as while deciding the welding sequences at the workshop. The other aspect that contributed to the failure was the presence of magnesium and chromium and somewhat higher percentage of carbon in the steel, which increased the possibility of notch brittleness in the high strength steel and also made the steel adjacent to the weld hard, and brittle on cooling, and consequently prone to crack development. In this case, lack of understanding in the engineering community about the implications of brittle fracture failures in welded structures caused the catastrophic failure.

Lessons learnt from this case study are the following:

- Proper care should be taken during design and detailing stage of welded steel structures.
- Special care, such as preheating and use of special electrodes, should be considered during fabrication stage, particularly for welding high strength steel material that may be prone to brittle fracture.
- Sequence of welding must be carefully examined before commencement of welding.
- Adequate inspection and supervision at every stage is an important aspect in the fabrication of welded steel structures.

There are also other instances of brittle fracture failures resulting in severing of complete sections of welded ships, pressure vessels, and pipelines. It is thus important that the designer is familiar with this phenomenon in order to eliminate the recurrence of such failures in future welded steel structures.

5.5 Concluding Remarks

Implications of fracture behavior of steel need to be properly understood by the designer. This is particularly significant while designing welded structure, because welding can considerably reduce the toughness of the material in locations near heat-affected zone. Careful design and detailing to avoid stress concentration, adequate quality control during fabrication, correct sequence of welding operations, as well as selection of appropriate steel material would certainly go a long way in preventing brittle fracture cracking and avoiding the possibility of failure of welded structures.

Bibliography

1. Hicks, J., 2001, *Welded Design—Theory and Practice*, Abington Publishing, England.
2. Hicks, J., 1999, *Welded Joint Design*, 3rd Edition, Industrial Press, New York.
3. Simpson, R., 1994, Brittle Fracture, in *Steel Designers' Manual*, 5th Edition, Knowles, P.R. and Dowling, P.J. (Eds.), Blackwell Scientific Publications, England.
4. Ghosh, U.K., 2006, *Design and Construction of Steel Bridges*, Taylor & Francis Group, London.
5. Francis, A.J., 1989, *Introducing Structures*, Ellis Harwood, Chichester.

6

Quality Control and Inspection

ABSTRACT The importance and necessity of quality control during fabrication are discussed. The concept of preparing an agreed quality assurance plan (QAP) to be followed in the fabrication stage is introduced. The relevant activities for achieving quality fabrication are dealt with in this chapter. The various methods of inspection (nondestructive and destructive tests) are discussed. The chapter ends with the salient features of inspection of a trial assembly of the fabricated structure to ensure that the same has been fabricated as per the requirements of drawings.

6.1 Introduction

In welded structures, quality control and inspection are of utmost importance in order to ensure that the completed structure strictly complies with the approved detail drawings and behaves in the manner envisaged at the design stage and achieves the desired durability. Normally, three major areas need to be covered while finalizing *quality assurance plan*, namely, material quality, weld quality, and dimensions. This plan needs to be tailor-made for the type of structural system adopted for a particular structure and should specifically address the areas that need special care during fabrication because of their sensitivity to imperfections. The frequency and intensity of inspection need to be of higher order in such cases. It should, however, be borne in mind that stringent inspection regime would certainly result in better fabrication quality, but, at the same time, would add to the cost of production. It is therefore necessary to make a balance between the user's necessity and the intensity of inspection. National codes and standards are available specifying the inspection procedures during fabrication and acceptable tolerances for the end product, which should generally be followed.

The concept of quality control involves broadly two agencies, namely, the fabricator's in-house quality control department and the client's/user's own quality control department or a nominated inspection agency. The fabricator is to follow the agreed QAP and the owner's inspection agency, in turn, is responsible for ensuring that the welded fabrication is done in accordance with the agreed QAP. It is also responsible for checking and accepting (or rejecting) the items offered for inspection by the fabricator. It is therefore important that only properly skilled persons are associated with this activity. Training of personnel on a continual basis is thus essential.

In welded fabrication, detail inspections take place at the end of each stage of welding process, results of which are recorded properly and communicated to the shop floor personnel for taking corrective actions as necessary. With this system, initial errors are not likely to be repeated, particularly in repetitive items. After completion of fabrication and trial assembly, the final inspection is done. During final inspection, emphasis is given on dimensional accuracy, open holes for site connections, and similar items to ensure proper alignment during site erection.

In this chapter, relevant aspects, which are generally considered to achieve quality fabrication of welded structures, are discussed.

6.2 Documentation

For achieving high level of quality of the final welded structure, it is imperative that documentation of all stage inspection activities is done properly. With this system, defective items or subassemblies cannot be used in the structure, unless these are rectified and passed by the inspection agency. Items/activities for which maintenance of record is recommended are as follows:

- Materials such as steel plates and sections and welding consumables
- Welding procedure
- Skill of welders and operators
- Layouts, templates, markings and jigs, and fixtures
- Weld preparation, fit-up, and assembly
- Dimensional tolerances in structures such as plate girders and open web girders
- Nondestructive and destructive inspection and tests
- Rejected items/subassemblies, including repair methods adopted

6.3 Materials

It is necessary to ensure that raw materials such as steel plates and sections, welding consumables, gas, and other bought out items are inspected before these are issued to the shop floor.

Steel plates and sections are to be checked for their physical and chemical properties by verifying the distinctive cast/heat marks stamped on each piece with corresponding numbers mentioned in the certificates issued by the producers. Over and above these, for important structures, independent sample tests are often carried out to confirm the accuracy of the mill test results. Test pieces for such additional tests should be cut out from the end of the material and processed for testing in the presence of the inspectors and duly documented.

Dimensional tolerances, that is, permissible variations from theoretical dimensions, of plates and sections from rolling mills, often present problems during fabrication. Also, materials may become twisted or bent during transit from the mills to the fabrication shops. These should be properly checked and rectified, as necessary, before being used in the job.

For welded fabrication, lamellar tearing often presents problems. Therefore, materials to be used for welding should be checked against presence of lamination. In particular, tests for lamination should be carried out for materials, in which tension stresses are transmitted through the thickness or where lamination could affect the buckling behavior of the member under compression. Lamination can be detected by nondestructive testing (NDT) method, such as ultrasonic examination, discussed later in this chapter. Extent of lamination should not exceed those indicated in relevant code of practice being followed.

As in the case of riveted or bolted fabrication, raw materials for welded structures should also be stored properly at the storage yard. These should be carefully unloaded, examined for defects, checked with documents, and stacked as per specification and size, above the ground on platforms, skids, or other suitable supports to ensure that these do not come in contact with water or ground moisture. Electrodes should also be stored specification wise and kept in dry, warm condition.

6.4 Welding Procedure

In order to ensure that acceptable quality of weld is achievable using the proposed welding procedure, it is necessary to examine the same before actual fabrication work is taken up. For this purpose, a specimen of welded joint of adequate length of the same cross section and material is made out with the same welding parameters (i.e., electrode, wire, flux, current, arc voltage, and speed of travel) that are proposed to be used in actual contraction. This joint is then subjected to required tests for acceptance of the welding procedure by the inspecting agency.

The welding procedure document should include the following items in respect of each type of welded joints:

- Sketches showing weld joint detail and welding parameters for each run.
- Detailed description (including sequence) of how the weld is to be done.
- Results of testing. The tests required for acceptance are specified in most national codes and these are to be followed in the workshops.

Having established the acceptable welding procedure for each joint, it should be ensured that the actual job is done in accordance with the same. In case there is a change in the welding position, weld preparation or fit-up conditions, a new specimen with the altered conditions should be made out and tested in the same manner as described earlier.

6.5 Skill of Welders and Operators

It is imperative that approval tests for each and every welder/welding operator is conducted in order to ensure his capability to produce welds of acceptable quality with the welding processes (manual, semiautomatic, or automatic), materials, consumables, and the procedures being used in the fabrication of the particular structure. Standard codes lay down the requirements that are to be followed for such tests, and welders and welding operators who have passed successfully in these tests are considered qualified to undertake the welding work involved in the particular structure. It must, however, be borne in mind that while the welders and operators are bound to pay special attention and effort during such tests, they may not do so under every production condition. Therefore, complete reliance should not be placed on these tests only. Quality of the welds produced on the actual structure under fabrication must also be checked during and after the welding.

6.6 Layouts, Templates, Markings, Jigs, and Fixtures

It is obligatory that dimensional checks of all full-scale layouts, templates, markings, jigs, and fixtures prepared in the template shops are subjected to dimensional checks on a routine basis. Mistakes detected at this stage are to be rectified before these are issued to the fabrication shop.

6.7 Weld Preparation, Fit-Up, and Assembly

Weld preparation and fit-up should be done in accordance with the standards laid down in the code of practice being followed. Needless to add, a correctly prepared and assembled joint will produce satisfactory results in the deposition of welds both in butt and fillet welds.

Dimensional accuracy of weld preparations is to be checked using a suitable gauge capable of measuring bevel angle, root face, and root gap (Figure 6.1). Preparation may be for simple flange to web fillet welded joints and in-line butt welded joints in plate girders, or for more complex cruciform joints and corner joints in box girders. It is expedient to check weld preparations of all critical joints, before welding work is undertaken. Also, inspection has to be carried out after the assembly of different components to ensure that the fit-up has been done as per orientation and measurements shown in the drawing.

6.8 Inspection Personnel

Fabricator's own inspection should be carried out by appropriately qualified and experienced personnel. In many cases, the purchaser engages the services of an independent agency to carry out inspection, in addition to the fabricator's inspection set up. Normally, this independent inspection agency posts an inspector or a team of inspectors to work on a full-time or part-time basis to oversee the inspection work and monitor the entire production process to ensure that the quality of this product is of acceptable standard.

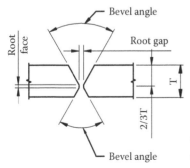

FIGURE 6.1
Typical asymmetric double-V butt welded joint showing bevel angle, root face, and root gap.

6.9 Inspection

6.9.1 General

Welded joints are to be properly inspected to ensure that the fabricated structures perform their intended functions satisfactorily. For a foolproof control of the quality of welded joints, it is expedient to have constant checkup at every stage of fabrication even before the welding starts. This method will provide opportunities for early detection of the flaws and introduction of corrective measures immediately, thereby avoiding possible serious defects in the structure after the fabrication has been completed. In fact, weld defects such as cracks, undercuts, undersize welds, and similar defects may prove to be rather expensive to rectify if detected too late.

In cases where the submerged arc welding process is adopted for forming the I-section, it is recommended to tack one run-on and one run-off piece at each end of the joint to be welded. These pieces are to be of the same material and thickness as the parent plates and prepared in identical manner, such that, the weld deposited on the run-on and run-off pieces form an integral part of the main weld. These may, then, be tested for the quality of the weld (see Chapter 9, Section 9.5).

Quality control in dimensions is an important aspect to be considered in welded steelwork. However, it is not possible to produce fabricated units to exact dimensions. Consequently, a batch of theoretically similar items cannot be produced exactly the same. In other words, it has to be accepted that some amount of dimensional deviations (from the drawings) will be present in any fabricated item. For this reason, standard codes allow certain amount of deviations from the required dimensions of fabricated parts. The limits of such deviations are termed *tolerances* and are clearly defined in specifications for workmanship. These deviations are normally taken into account in the factor of safety while computing the member capacity. Imperfections beyond these limits should be rectified before further processing of the deficient workpiece.

The methods commonly employed for inspection can be broadly divided into two categories, namely, nondestructive and destructive inspection and tests.

6.9.2 Nondestructive Inspection and Tests

In nondestructive inspection and tests, the inadequacies and defects in the welded joints are detected without impairing the quality and functions of the joint. Most defects in welds such as cracks, inclusions, lack of penetration, lack of fusion, undercut, and porosity can be evaluated by nondestructive inspections and tests. The following are the common methods:

- Visual inspection
- Liquid penetrant testing
- Magnetic particle inspection (usually abbreviated as MPI)
- Radiographic test
- Ultrasonic test

6.9.2.1 Visual Inspection

Visual inspection can furnish a good indication of the quality of the welds, provided the inspector is competent and has reasonably good power of observation. This inspection is the simplest and most widely used inspection method. It is fast, economical, and does not require any sophisticated equipment. In fact, good welds can be recognized from their color, shape, size, and general appearance. Equipment needed are inspection gauge, magnifying glass, mirror, scale, caliper, and so on. In general, visual inspection is carried out in three stages:

1. Visual inspection prior to welding
   In effect, this is a preventive inspection, and can eliminate conditions that may lead to serious weld defects. This activity involves inspection of the following items:
   a. *Parent metal:* Dimensions of all plates and sections should be checked. Verification of the quality is to be carried out by referring to the corresponding test certificates. If considered necessary, spot checks on the chemical composition and physical properties may be made by cutting test specimens from the parent metal. It should also be ensured that the metal is free from surface flaws, lamination, and imperfect edges.
   b. *Welder's qualification:* Welders and welding operators who are qualified in accordance with the qualification tests should only be permitted to work.
   c. *Welding consumables:* These should be checked for quality. Electrodes should be free from moisture.
   d. *Welding equipment:* This should be checked for workability in accordance with the approved welding procedures.
   e. *Assembled unit:* After the components are assembled in position for welding, these should be checked for root gap; root-face dimension; edge preparation; dimensions of parts; finish; and fit-up, including back strips, alignment, and cleanliness.
   f. *Welding procedure:* Prior to actual welding, it should be ensured that the procedure to be adopted for the particular job is in accordance with the approved welding procedure.

2. Visual inspection during welding

Visual inspection needs to be carried out while welding is in progress to ensure that the approved procedure is strictly followed throughout the welding activity. The items to be checked include the following:

a. Preheating

b. Maintenance of welding parameters, namely, current, arc voltage, and speed of travel

c. Sequence of welding

d. Cleaning of slag after each run in multi-run welding

e. Proper fusion

f. Electrode spattering

Also, consumables should be inspected periodically to ensure that the specified quality is maintained consistently. Another important area for checking is the weld run, which will become inaccessible after subsequent welds.

3. Visual inspection after welding

Prior to carrying out visual inspection after welding, it is necessary that the weld surface is thoroughly cleaned of oxide layers and adherent slag. Chipping hammers should be avoided for this activity since the hammer marks are likely to obscure the evidence of fine cracks. Brushing with stiff wire brush or grit blasting may be adopted for this purpose.

The items to be checked include the following:

a. *Weld profile:* The final weld profile should conform to the size and contour recommended in the standard specifications. For this purpose, welding gauges are available.

b. *Surface defects:* These include unfilled craters, lack of fusion crack, porosity, undercut, and slag inclusion.

c. *Weld appearance:* These include irregular ripple marks, peening marks, and surface roughness.

d. *Disposition of welds:* This should be as shown in detail drawings.

e. *Dimensional accuracy of fabricated items:* Distortions and warping should be within allowable tolerance limits.

Although most of the visual inspections are done with unaided eyes, use of optical instruments may be beneficial in some specified areas, which cannot be easily seen or are not accessible to the unaided eyes. The use of optical microscopes, borescopes, endoscopes, and telescopes is recommended in such cases.

6.9.2.2 *Liquid Penetrant Testing*

This method of testing is carried out for detecting certain flaws open to the surface such as cracks and surface porosities. First, the weld surface is cleaned by removing scale, rust, grease, and paint and dried. Then the penetrant is applied uniformly by spraying to form a film over the surface. The penetrant is drawn into any cracks or other defects by capillary action. After the penetration time (as recommended by the suppliers), the excess penetrant on the surface is cleaned off either by an appropriate solvent or by washing with water. A developer such as chalk powder (dry or suspended in a liquid) of contrasting color (usually white) with high absorbent property is then applied. If any cracks or surface defects are present, the penetrant bleeds out of the defects by reverse capillary action and appears as a stain on the surface, which is broader than the actual flaw and is much more visible.

Liquid penetrant can be either a dye penetrant, which is commonly of red color, or a fluorescent penetrant. In the former case, red color stain contrasts sharply with the white background of the developer. The fluorescent penetrant, on the other hand, is more prominently visible with a brilliant glow under ultraviolet light and consequently more sensitive compared to dye penetrant. This method, however, requires a dark room and a source of ultraviolet light. For this drawback, usually color contrast dye penetrant is used in preference to fluorescent penetrant.

6.9.2.3 *Magnetic Particle Inspection*

MPI can be used for detecting surface and subsurface discontinuities (for both butt and fillet welds), such as cracks, porosities, slag inclusions, fusion deficiencies, undercuts, incomplete penetration, and gas pockets. This method can be gainfully utilized where radiographic or ultrasonic methods are not available or are not practicable due to constraints arising from shape or location of weldment.

In this method, a magnetic field is first set up within the piece to be tested and then fine iron particles (either dry or suspended in liquid) are dusted (or sprayed) over the test area. If there is a crack, the magnetic field will be distorted and a small north–south pole area at the crack zone will be created and the outline of the crack will be clearly visible. Generally, surface cracks are indicated by a sharp line of magnetic particles following the crack line. In case of subsurface cracks, somewhat indistinct collection of magnetic particles are shown on the surface.

The magnetic field in the test area may be created by adopting a number of methods. In the first method, the poles of a strong permanent magnet or an electromagnet are placed on either side of the area to be tested. In the second method, electric current is passed through the test piece itself. In the third method, magnetic field is created by induction by winding a coil around the test area and then passing current through the coil.

6.9.2.4 Radiographic Test

This method is suitable for testing welded joints, which have access from both sides of the test area, with the radiation source placed on one side and the film on the other side. The system essentially consists of passing x-rays or gamma rays through the weld being tested and creating an image on a photosensitive film, placed in a cassette or holder. If there are voids in the weld, less radiation will be absorbed by the steel and consequently more radiation would pass through that area to the film. Defects will appear on the film as areas darker than the image produced by the surrounding areas of defect-free material. Similarly, inclusions of low density, such as slag will appear as dark areas, whereas inclusions of high density (e.g., tungsten) will appear as light areas. The method is useful for detecting both surface and subsurface defects in welding such as cracks, inclusions, porosities, fusion deficiencies, incomplete penetrations, and undercuts. No doubt, this is an expensive method of testing compared to other NDT methods. Also, the method requires specialized knowledge and considerable professional judgment in selecting the angles of exposure, as well as interpreting the results recorded on films. Another aspect that needs special mention here is safety hazard posed by radiation. It is necessary to abide by certain statutory safety requirements for transporting and storage of radioactive isotopes used in the system, as also in radiographing the weld so as to avoid radiation exposure to human body. In spite of all these shortcomings, however, radiography is considered to be a preferred testing method over other methods, particularly for important structures as it has got a big advantage of providing a permanent record of every test carried out.

6.9.2.5 Ultrasonic Test

Ultrasonic testing is a portable and yet a highly sensitive method for detection of surface and subsurface flaws in the welds (such as cracks, porosity, and lack of penetration), as well as defects in the parent metal. It is widely used as an important tool for quality control in welded steel fabrication industry. In this process high-frequency beams of sound waves of short wavelengths are introduced by means of an ultrasonic transducer into the area to be tested. The sound beams travel in a straight line through the steel. The particular beam, which is interrupted by a discontinuity (defect), reflects back to the transducer. This produces a voltage impulse, which appears on the cathode ray tube oscilloscope. This is a highly sensitive testing system and requires a well-trained and experienced operator with considerable skill for correctly interpreting the pulse-echo patterns appearing on the screen. This system has an advantage, in that, it requires access from only one side of the material, unlike radiography, where access from both sides is required. However, unlike radiography, this method does not provide any permanent test records.

6.9.3 Destructive Tests

Destructive tests are generally conducted for approval of welding procedures, qualification tests for welders and also for quality control during production process. These tests can be broadly divided into these categories:

- Chemical analysis
- Metallographic testing
- Mechanical testing

6.9.3.1 Chemical Analysis

Chemical analysis is primarily done to ascertain the chemical composition of the solidified weld metal after the welding is completed. This weld metal is essentially a mixture of the molten parent metal and steel from the electrode. During welding, many elements get evaporated or lost in the process, necessitating the chemical analysis to confirm that the composition of the solidified weld is not inferior in quality than and also compatible with the parent metal.

This exercise is particularly important if a structure is situated in a corrosive environment, where the weld metal may be more susceptible to corrosion than the parent metal.

6.9.3.2 Metallographic Testing

Metallographic testing is often necessary to examine the soundness of the final welded joint in view of certain changes occurring in the metallurgical structure of the steel in the fusion zone and the heat-affected zone as the joint cools. The primary aim for the tests (macro-etch and micro-etch tests) is to examine the distribution of the nonmetallic inclusions, metallurgical structure, depth of penetration, quality of fusion, and so on. In practice, a section of the weld joint is cut and the specimen is prepared as per the standard metallographic practice. The macro-etch testing is conducted by etching the specimen to bring out the macrostructure and examining by unaided eye or at low magnification of the order of 30 times. In micro-etch testing, on the other hand, the specimen is polished, etched, and examined under high magnification (more than 50 times) using optical or electron microscope.

6.9.3.3 Mechanical Testing

The purpose of mechanical testing is to determine the mechanical properties of the final weld, such as tensile strength and ductility. Specimens may be cut by sawing or by thermal process. In the latter case, the materials at the edges are likely to be discarded while preparing the test piece. Some of the common tests recommended by different authorities are briefly discussed in the following paragraphs.

6.9.3.3.1 Longitudinal Tensile Test

This test is carried out to determine the tensile strength of a butt welded joint in the longitudinal direction of the weld. The specimen is prepared from the weld metal only as shown in Figure 6.2.

The specimen is ruptured under tension load to determine the breaking strength, percentage elongation, percentage reduction in area, as well as the type and location of any flaw on the fractured surface.

6.9.3.3.2 Transverse Tensile Test

In this test, the tensile strength of a butt welded joint in the direction transverse to the weldment is determined. The specimen is prepared in accordance with the governing code and subjected to tensile test to determine breaking strength, as in the case of longitudinal tensile test stated in the previous paragraph. A typical test specimen is shown in Figure 6.3.

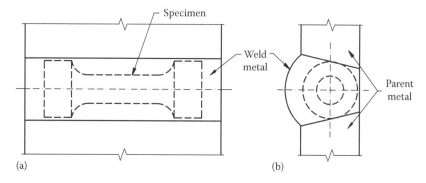

FIGURE 6.2
(a, b) Preparation of specimen for longitudinal tensile test of weld metal.

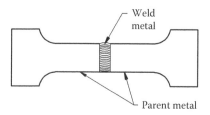

FIGURE 6.3
Transverse tensile test specimen.

6.9.3.3.3 *Hardness Test*

Hardness measurement is relevant in welded fabrication work for determining whether there has been any significant change in the property of steel in the welded joint area by heat generated during welding operation. The commonly used hardness testing methods are as follows:

- Brinell hardness test
- Rockwell hardness test
- Vickers hardness test

Brinell method can be used for hardness survey over a relatively large area, such as face of weld and the parent metal. In this method, the specimen is mounted on the testing machine and a hardened steel ball of specified diameter is placed on it. A specified load is then applied on the hardened steel ball for about 30 seconds and then released. The size of the depression (indent) on the specimen is then measured and the corresponding Brinell hardness number is obtained by an empirical formula.

The basic principle of Rockwell method is similar to Brinell method. It differs from the former in that a lesser load is applied on a smaller ball or a cone-shaped diamond. The size of the indent is measured and shown in a dial attached to the machine. As in the case of Brinell method, the hardness is expressed as Rockwell numbers.

Vickers hardness test was developed in England in the early twentieth century and was known as the *diamond pyramid hardness test*. In this case, the indenter is of pyramid shape. It is a rather sophisticated system requiring the test point to be highly finished to enable an optical device to measure the size of the indent accurately enough to give useful result.

Both the Rockwell and Vickers methods can be used for hardness measurement over only small zones such as weld cross section, heat-affected zone, and individual weld bead.

6.9.3.3.4 *Guided Bend Test*

This test requires inexpensive equipment and is widely used in fabrication industry for ascertaining soundness and ductility of the weld metal at the face and root of a butt welded joint. Essentially, it consists of bending the specimen 180 degrees around a pin of about 40 mm diameter by hydraulic jack or other device. The test specimen is bent in such a diameter that the weld tends to pull away from the two edges of the joint. The outside surface elongates due to tension, while the inside surface shortens due to compression, and the test piece bends without failure. Figure 6.4 illustrates typical shapes of the test piece after the bending.

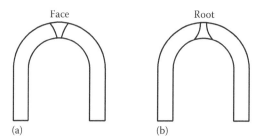

FIGURE 6.4
Specimens for guided bend test after bending: (a) face bend and (b) root bend.

6.9.3.3.5 Free Bend Test

This is a variant method guided bend test, which also indicates the ductility of weld metal in a butt welded joint.

6.9.3.3.6 Nick-Break Test

This test is carried out to examine possible internal defects in a butt welded joint, such as porosity, slag inclusion, degree of fusion, and ductility. The test specimen is made from a butt welded joint by cutting out a section of specified dimensions. Each edge of the weld of the joint is nicked by means of a suitable instrument (e.g., saw) through the center (Figure 6.5). When the specimen is placed across two steel supports, and struck with a heavy hammer, it will break at the nicks. The weld metal is now exposed for examination for defects in accordance with the standard being followed.

6.9.3.3.7 Fillet Weld Break Test

This test is carried out to determine the soundness of fillet welds, such as, porosity, slag inclusion, and root fusion. A test piece of fillet welded joint

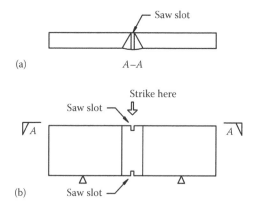

FIGURE 6.5
(a, b) Nick break test.

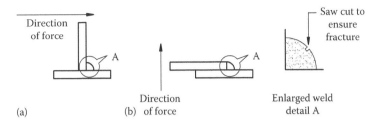

FIGURE 6.6
Fillet weld break test specimens: (a) Tee joint and (b) lap joint.

of specific details such as Tee joint and lap joint with saw cut at the middle (Figure 6.6) is prepared and subjected to force in a specific direction by means of a press, a testing machine, or blows of hammer until the fillet weld fractures. The fractured surface is then examined for soundness.

6.9.3.3.8 Shear Strength Test for Fillet Weld

This test is preformed to ascertain the shear strength of a fillet welded joint. The test is simple and inexpensive and is commonly used in workshops for periodic check on the quality of the fillet welds. Tests are carried out separately for longitudinal and transverse welds by preparing fillet weld test pieces to details as per standard being followed and subjecting them to tensile load to break or fracture the joints.

6.9.3.3.9 Notched Bar Impact Test

This test is carried out to ascertain the impact strength of weld metal. This strength is primarily dependent on heat, chemical composition, and presence of certain embrittling elements in the weld metal. Consequently, weldment, which is ductile under uniform static loading at normal temperature, may show marked reduction in its ductility under impact loading at low temperature.

Typically, a V-notched bar test piece of specified dimension is used in the test. Two types of tests are available, namely, Charpy and Izod. In both types, the test is done by breaking the test piece by blow from a swinging hammer in the vertical plane. In the Charpy impact test, the test specimen with a notch at the center is supported as a simple beam at both ends placed on a split anvil. The notch is on the vertical face away from the point of impact. The test piece is broken by a blow on the face opposite to the notch. In the Izod test, the bar is held vertically in a vise, with the notch placed just above the jaws, and broken by a blow applied on the side of the notch. Figure 6.7 shows a schematic arrangement of these tests.

In both the cases, the heavy pendulum hammer falling through a fixed distance strikes the specimen at a fixed velocity, and slows down and swings

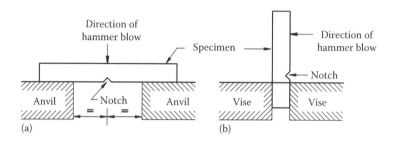

FIGURE 6.7
Schematic arrangement of impact tests: (a) plan view Charpy test and (b) elevation Izod test.

forward to a lower height as the test piece absorbs some energy. The difference of the height to which the hammer rises after striking the test piece and the height from which it started is measured. From this difference and the effective weight of the hammer, the energy absorbed by the test piece is determined.

6.9.4 Inspection of Trial Assembly

It is expedient to erect all important structures in entirety in the workshop itself to prove that the final structure has been fabricated as per the requirements of the drawings. This mock up assembly is termed *trial assembly*. Inspection of the trial assembly is the final inspection prior to application of primer paint and dispatch of components to site for erection.

The final inspection is normally carried out jointly by the fabricator's and the client's inspection teams. The main items to be inspected are as follows:

1. Dimensional accuracies
   a. Overall length
   b. Length between supports
   c. Lengths of individual panels
   d. Height of the structure
   e. Distance between centers of trusses and girders
   f. Squareness in plan
   g. Verticality
   h. Camber, if necessary
   i. Deflection
2. Connections
   a. Edge preparations for field welding (accuracy of shape, size, gap, etc.)
   b. Accuracy of drilling in case of field connections by bolts or rivets

6.10 Concluding Remarks

The activities required for quality control and the inspection regimes described in the foregoing sections would certainly add additional costs to the job, directly in themselves, and indirectly by time delays due to interruptions in the production schedule. However, considering the overall safety aspect, particularly in welded structures subjected to impact load or fatigue conditions, a good case is made for such a QAP. However, in all cases, the importance of the structure should dictate the type of inspection and acceptance criteria to be adopted. The controlling standard specifications generally provide the guidelines to minimize the additional cost.

Bibliography

1. Hicks, J., 2001, *Welded Design—Theory and Practice*, Abington Publishing, England.
2. Gourd, L.M., 1995, *Principles of Welding Technology*, Edward Arnold, London.
3. Blodget, O.W., 2002, *Design of Welded Structures*, The James F. Lincoln Arc Welding Foundation, Cleveland, OH.
4. Ghosh, U.K., 2006, *Design and Construction of Steel Bridges*, Taylor & Francis Group, London.
5. Evans, J.E., Iles, D.C., 1998, *Guidance Notes on Best Practice in Steel Bridge Construction*, The Steel Construction Institute, Ascot.
6. Samanta, A.K., 2009, *Guidebook for Fabrication with Steel*, Institute for Steel Development & Growth, Kolkata, India.

7

Design Considerations for Welded Joints

ABSTRACT   This chapter deals with some of the practical aspects of fabrication, erection, and service performance of welded structures, which should be considered by the designer. Initial planning of layout, locations of joints, and make up of sections, as well as weldability aspects of the proposed steel grade, have been discussed. Other relevant areas deliberated are load conditions, selection of joint types, weld size vis-à-vis cost, distortion, choice of edge preparation, and ease of fabrication and erection.

7.1 Introduction

Design in any form of construction is effective only when the practical aspects of fabrication and construction are carefully considered during design stage itself along with the structural functions. This is certainly true in the case of an efficient and economic welded structure.

Some of the practical points that require consideration for design of welded structures are dealt with in this chapter. Careful study of these points is advised for those designers who have not been adequately exposed to the practical side of the subject.

7.2 Layout, Locations of Joints, and Make Up of Sections

In cases where components are to be formed by welding individual plates, the layout of the structure and locations of the joints should be planned in the preliminary design stage to avoid wastage of materials. In this context, the designer should consider use of standard rolled sections, such as beams and channels, rather than forming individual parts by welding plates, as the former option is likely to reduce costs without sacrificing quality.

7.3 Weldability of the Material

For any welded structure, the material must be of a quality that when welded will perform satisfactorily in the service condition. Important aspects that need particular attention are the chemical composition and carbon equivalent of the material, its ductility, fracture toughness, and strength when welded. These aspects play an important role in the susceptibility of the welded joints to fracture, and therefore, stability of the entire structure. Thus, selection of the material is a very important aspect in the design of an efficient welded structure. These aspects are discussed in Chapter 2. In this connection, one point needs special mention. Although theoretically steel is homogeneous, having equal properties in all directions, in practice, it is not so. In fact, many properties of rolled plates are directional; that is, some desirable properties are in the direction of the rolling. Thus, strength and ductility are often quite low across the thickness due to the presence of nonmetallic impurities during rolling. Loading plates in the direction of through thickness may cause lamellar tearing due to weld shrinkage, and should be avoided (see Chapter 3).

7.4 Load Conditions

The designer needs to identify clearly the different static load conditions that the welded structure will be required to withstand during service conditions, as well as during the erection stage, and analyze the structure accordingly. Locations of critical stresses (e.g., tension, compression, bending, shear, and torsion stresses) are to be determined and the structure is to be designed to satisfy these stresses. Wherever possible, joints should be located away from high stress areas.

In structures that are likely to be subjected to fatigue conditions during service life (as in a crane girder), the effects of fatigue need to be examined. The structural components as well as the joints are to be designed accordingly. Fatigue in welded structure is discussed in Chapter 9.

Also, the phenomenon of brittle fracture is an important aspect in the design of welded structures and should be considered at the design stage. This topic is discussed in Chapter 5.

Corrosion resistance is another aspect that needs due attention at the design stage, and necessary provision should be incorporated in design and detail.

7.5 Joint Types

The types of joints commonly used in welded structures are lap joints, butt joints, corner joints, and T-joints. The type of joint to be used in a particular location is generally governed by the nature of the components to be joined. At the same time, each welding process used in fabrication has its own characteristics and capabilities. In this connection, access to the location of welding and subsequent inspection will play an important role in the final selection of the welding process. Thus, the particular joint type and its dimensions need to be suitable for the desired welding process. In practice, the choice is narrow, since it ultimately depends on the process actually available with the fabricator.

7.6 Weld Types

As already discussed, welded joints are made by using two main types of welds: butt weld or fillet weld. These two types are not just different in form, but they represent two different engineering philosophies. In the case of butt weld, two pieces are joined together mostly by fusing their complete cross sections, whereas fillet weld connects the two pieces outside their cross section with lines of weld metal as necessary. Of these two types, butt welded joints are generally considered to have the potential for better performance (e.g., in case of fatigue conditions), but it can be costlier to make.

7.7 Weld Size

Aspects to be considered while specifying weld sizes are discussed in the following paragraphs.

7.7.1 Cost

Welds should be designed to match (not exceed) the required strength of the joints and the codal requirements. Unnecessary over-welding only increases the cost without adding to the strength of the joint. The designer should consider the fact that the strength of a fillet weld is proportional to its size, while the volume and weight of the weld are proportional to the square of the size.

7.7.2 Residual Stresses and Distortion

Welding sometimes induces residual stresses from weld shrinkage, leading to distortions in the structure. Larger the weld size, greater would be the magnitude of distortion. This topic is discussed in Chapter 4.

The designer should, therefore, ensure not to specify weld sizes larger than the governing code demands, lest this would add to the cost, as well as to the extent of distortion.

7.8 Edge Preparations

While thin plates can be butt welded with square edges, it is a common practice to shape or prepare the edges of thicker plates prior to butt welding operation. The primary reason of this edge preparation is to permit accessibility to all parts of the joint during welding in order to ensure adequate fusion throughout the cross section of the weld. Recommended types of edge preparations for various thicknesses of materials are available in the code of practice followed in the design. However, for the choice of the particular type to be adopted, the designer needs to consider the following aspects:

- Configuration of the joint
- Access for welding and inspection
- Availability and cost of the cutting equipment
- Proneness to distortion

A typical configuration of a simple single-bevel butt joint is shown in Figure 7.1, which defines the nomenclatures such as included angle, angle of bevel, root face, and root gap, commonly used in the industry. The edge preparation shown in this joint can be made cheaply by gas cutting. The main purpose of the root face is to provide an additional thickness of metal, in order to minimize any *burn through* tendency in case of a sharp edge (or feather edge), and to ensure a consistent run of weld. Root faces are, however, not recommended where the butt weld is made on to a backing strip to avoid formation of gas pocket between the weld and the backing strip. In such cases, sharp or feather edges are preferable, where the edge may be fused into the back up strip. As regards cost of preparation, a feather edge is generally made by one cut by the torch, while a root face needs at least two cuts by the torch or a torch cut plus machining.

For butt welds in thicker materials, it may be desirable to adopt a two-sided bevel, rather than a single-sided one. Figure 7.2 illustrates two typical double-V preparations. These configurations have two advantages. First, for identical thicknesses, a joint with single-V preparations consumes double the weld

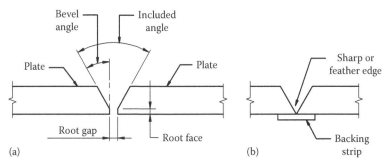

FIGURE 7.1
Single-V preparations: with (a) root face and (b) sharp or feather edge.

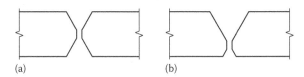

FIGURE 7.2
Double-V preparations: (a) symmetrical and (b) asymmetrical.

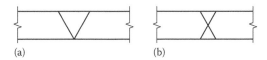

FIGURE 7.3
Comparison of volumes of weld metal: in (a) single-V and (b) double-V weld preparations.

metal than that, if double-V preparations are used (Figure 7.3). Consequently, the cost of welding would be reduced in the latter case. Second, the extent of distortion is much reduced in the joint if double-V preparations are used. This is because of balanced heat input through the thickness during the sequence of welding operation. To minimize distortion further, often, asymmetrical preparations are made. See Figure 7.2b. The double bevel with a root face is commonly gas cut quite cheaply in one pass using three multiple burning torches (Figure 7.4). However, for butt welding workpiece involving double-V preparations, the joint may have to be turned over to weld the second side. This would entail additional work during fabrication leading to additional cost. This aspect needs consideration during the selection of the type of joint configuration.

Another option of edge preparation is the J or U preparation (Figure 7.5). While this type of preparation offers even distribution of weld metal throughout the depth of the joint and consequently is likely to reduce

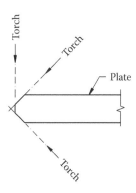

FIGURE 7.4
Cutting of a plate edge for double-V weld preparation.

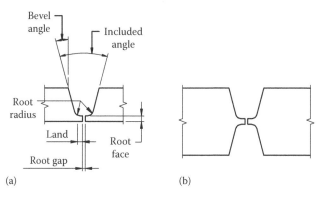

FIGURE 7.5
U preparations: (a) single-U weld and (b) double-U weld.

the extent of distortion, it requires elaborate machining as opposed to simple gas cutting required for V preparations. This type of preparation is consequently not very popular to the fabricators, and is used only on special demands.

7.9 Ease of Fabrication and Inspection

It is necessary for the designer to become familiar with the workshop conditions and the challenges that are likely to be faced and overcome in translating drawings into fabricated products. It has often been found that the joint specified by designers cannot be easily welded because of lack of access, which, for manual welding, means allowing the welder not only to

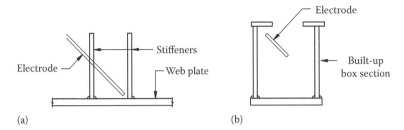

FIGURE 7.6
Lack of access for welding: (a) electrode not reachable at correct angle and (b) welder not able to see.

see the joint, but also reach it with the welding rod at the correct angle and at the same time being able to see the arc and effectively manipulate it along the joints. Figure 7.6 illustrates typical details with lack of access for welding. In case of automated welding facilities also, individual system has its own requirement of space around the joint, which has to be considered by the designer while adopting the joint type. Often, insufficient space for welding heads of automated system for welding stiffener to a plate girder makes welding difficult. It should be understood that requirement of the particular facilities available with the fabricator may often dictate the type of welded joint detail to be adopted in the structure.

The main positions of welding are illustrated in Figure 7.7. Out of these, the flat (downhand) position is the most advantageous for welding due to the following reasons:

- Maximum deposit rate of weld metal
- Maximum speed of welding
- Least cost of welding
- Ease of welding and least fatigue of welders

Manipulating devices are often used to rotate the work, so that welds are made in the flat position. A typical manipulator for welding of plate girder is shown in Figure 7.8. Feasibility of inspection is another factor, which needs to be considered for the choice of the joint and weld. Inspection may involve various means ranging from visual surface examination, assisted visual inspection, such as magnetic particle and dye penetrant, to radiography and ultrasonic test (Chapter 6), which have their individual requirements for access.

The foregoing aspects need to be taken into consideration when designing the joint and the weld. Otherwise, a theoretically elegant design may turn out to be impractical as this cannot be fabricated easily and cost effectively.

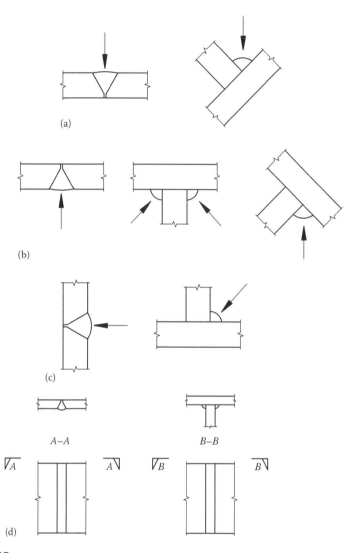

FIGURE 7.7
Welding positions: (a) flat (downhand), (b) overhead, (c) horizontal–vertical, and (d) vertical positions.

This situation should be guarded against. Also, for large structures, it is sometimes preferable to fabricate subassemblies of structures first, which could then be finally assembled and welded. Subassemblies are light and easier to handle. This system also allows distortion control at intermediate stages of fabrication.

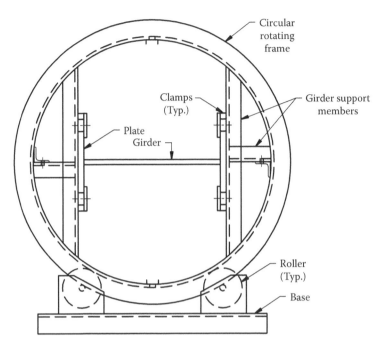

FIGURE 7.8
Manipulator for plate girder fabrication.

7.10 Concluding Remarks

As in any other fields of engineering design, it is not easy to find a unique "correct" solution in the design of welded steel structures as well. The discussions on the various considerations embodied in the preceding paragraphs may, in some cases, be conflicting between each other; the designer, in such cases, may have to make compromises to arrive at the most suitable and appropriate solution for a particular problem.

Bibliography

1. Hicks, J., 2001, *Welded Design—Theory and Practice*, Abington Publishing, UK.
2. Hicks, J., 1999, *Welded Joint Design*, 3rd Edition, Industrial Press, New York.
3. Blodgett, O.W., 2002, *Design of Welded Structures*, The James F. Lincoln Arc Welding Foundation, Cleveland, OH.

8

Design of Welded Joints

ABSTRACT Guidelines for design of butt weld and fillet weld are covered. Strength of full penetration and partial penetration butt welds, effective length, and limitations of intermittent butt welds are discussed. Types of fillet welds, methods for computation of size, effective throat thickness, effective length, and strength as also step-by-step procedure for design of fillet welded joints have been deliberated in detail. Other topics related to fillet weld covered are end return, lap joints, combined stresses, bending and compression stresses, and intermittent fillet welds. Analyses of two typical fillet welded eccentric connections are provided at the end.

8.1 Introduction

Most commonly used weld type in steel structures is fillet weld, which roughly accounts for 80% of the welded joints used. Butt weld occupies the second place in popularity, with about 15% of the welded joints. The remaining 5% comprises slot, plug, and resistance welding.

Guidelines for the design of the two major types of welds are discussed in the present text.

8.2 Butt Weld

As defined earlier, a butt welded joint is made within the surface profile of the joining members, either by fusing the square butting surfaces of the components or by filling a gap (produced by preparing the edges) by molten weld metal. Computation of strength of different types of butt welds is discussed here.

8.2.1 Full Penetration Butt Weld

The strength of a full penetration butt welded joint is, in effect, the minimum of the strength of the weld metal, the heat-affected zone, and the parent metal, which, collectively, constitute the cross section of the joint. The selection of

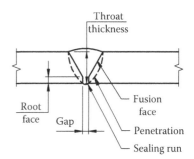

FIGURE 8.1
Full penetration butt weld.

electrode and welding parameters (current, speed, voltage, etc.) play a vital role in developing the strength of a butt welded joint. The capacity of the butt weld is based on its throat thickness (see Figure 8.1). In general, weld metal is made stronger than the parent metal. Also, the heat-affected zone is normally of adequate strength compared to the parent metal. Thus, in the case of a full penetration butt weld joining two plates of identical quality and thickness, the joint is considered adequate to transmit the force in the parent metal. For plates of dissimilar quality, the capacity of the weaker part is to be considered for full strength.

In case of high strength steel, it may not be always possible to obtain weld metal and/or heat-affected zone to match the high strength of the parent metal. In such cases, due consideration for reduction in the design strength based on the parent metal strength may be necessary. The preferred shape and dimensions of edge preparations for different thicknesses of plates are given in the national codes of practice, and these should be followed.

In order to achieve full penetration, certain aspects in the welding procedure need to be addressed during fabrication stage. In case of butt weld from one side of the workpiece, backing strip should be provided, and root faces must be melted to ensure good penetration. (Figure 8.2). When butt weld is

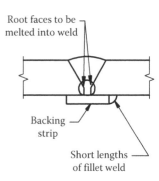

FIGURE 8.2
Backing strip.

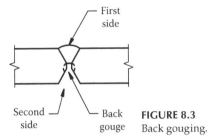

Second side — Back gouge

FIGURE 8.3
Back gouging.

done from both sides of the workpiece, the first (root) run on the firstside should be gouged on the second side and then the resultant cavity should be filled by the first run on the second side (Figure 8.3). It is necessary that the root faces are kept to the minimum; otherwise a large amount of metal will be left for gouging out. This will entail extra cost, as well as the risk of excessive distortion.

8.2.2 Partial Penetration Butt Weld

For partial penetration butt weld, the depth of preparation is considered to be the depth of penetration or throat thickness. For V butt weld, some national standards recommend a deduction of a fixed dimension, say 3 mm (or a proportion of the penetration) from the idealized throat thickness to obtain the net throat thickness for the purpose of calculating the tensile and shear capacity of the joint. This is done to cater for possible lack of fusion at the root (see Figure 8.4). For calculating the capacity for compression, the depth of root may be added to the throat thickness, provided the root faces are closely butted before welding. In single-V butt weld, the secondary bending due to eccentricity of the throat area needs to be considered in

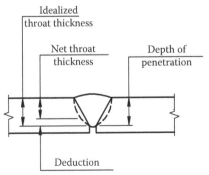

FIGURE 8.4
Partial penetration butt weld.

the calculation. Generally, partial penetration butt weld should not be subjected to tensile force or bending moments along the longitudinal axis of the welds. Also, this type of weld may give rise to high stress concentration under alternating or impact loads and are therefore not recommended for such conditions.

8.2.3 Effective Length

Effective length is the length of the butt weld, which is considered while computing the capacity or strength of a butt weld. Some allowance is normally made for possible lack of adequate throat thickness and/or presence of craters at the commencement and at the end of the butt weld. In order to ensure that the entire length of the main weld is of required size and is free from craters, it is a common practice to provide run-on and run-off plates at the two ends of the butt weld (see Chapter 9, Section 9.5).

8.2.4 Intermittent Butt Weld

Intermittent butt weld is not recommended for use in locations subjected to dynamic, repetitive, and alternating stresses, such as crane and wind girders. In fact, use of this type of butt weld is not very common in structures. However, in case it is unavoidable, the following conditions should be adhered:

- The weld should have an effective length of not less than four times the weld size.
- The longitudinal clear spacing between the effective lengths of successive welds should not be more than 16 times the thickness of the thinner part joined.

8.3 Fillet Weld

Unlike butt joint, a fillet welded joint is done by depositing weld metal outside the profile of the joining components.

8.3.1 Types of Fillet Welds

8.3.1.1 Normal Fillet Weld

In this type of fillet weld, the depth of penetration beyond the root is less than 2.4 mm.

8.3.1.2 Deep Penetration Fillet Weld

For deep penetration fillet weld, the depth of penetration beyond the root is 2.4 mm or more.

8.3.2 Size of Fillet Weld

Size of a normal fillet weld is defined as the length of the shorter side (leg length) of a triangle inscribed within a fillet weld as shown in Figure 8.5.

For deep penetration fillet weld, where the minimum depth of penetration beyond the root is 2.4 mm, the size of the fillet should be taken as the minimum leg length plus 2.4 mm (Figure 8.6). In case of fillet welds made by semiautomatic or automatic processes, the depth of penetration is considerably in excess of 2.4 mm. Some authorities allow the actual depth of penetration of such welds to be considered while computing the size, subject to agreement between the concerned parties.

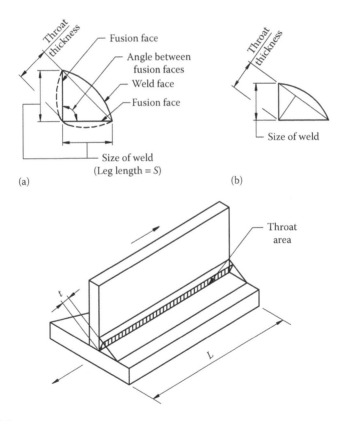

FIGURE 8.5

Typical fillet weld detail: (a) fillet of equal leg length and (b) fillet of unequal leg length—*t* is the throat thickness, *L* is the effective length, *S* is the size of weld, and *L* multiplied by *t* is the throat area.

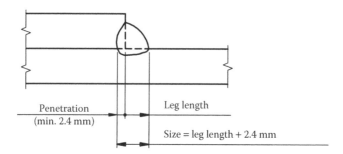

FIGURE 8.6
Deep penetration fillet weld.

Most authorities specify that the size of a fillet weld should not be less than 3 mm. The minimum size corresponding to the thickness of the component to be welded should be adopted as per recommendations of the code being followed.

Also, the minimum size of the first run or of a single-run fillet weld depends on the thickness of the thicker plate, to ensure proper fusion of the weld to the part and to minimize the chance of brittle fracture due to rapid cooling. The thicker part (particularly when the thickness is more than 50 mm), should be adequately preheated to prevent cracking of the weld. The minimum size of run of a fillet weld should be as recommended by the guiding code.

Unlike the minimum size of fillet weld, the maximum size is normally a function of the thickness of the thinner part joined. For example, where fillet weld is required at the square edges of a part, most authorities recommend that the maximum size of the weld should be limited to 1.5 mm less than the thickness of the part (Figure 8.7). This detail would reduce the chance of melting of the exposed area of the parent metal and ensure development of the total strength without overstressing the adjacent metal. However, where the above stipulation cannot be achieved and

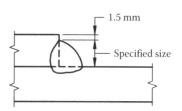

FIGURE 8.7
Maximum size of fillet weld.

FIGURE 8.8
Unacceptable fillet weld detail because of reduced throat thickness.

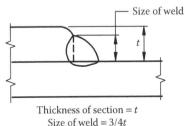

Size of weld

Thickness of section = *t*
Size of weld = 3/4*t*

FIGURE 8.9
Fillet weld at rounded toe of rolled section.

the specified size of the weld is such that it builds up to the full thickness of the parent metal, special care must be taken to ensure that no melting of the parent metal takes place, thereby reducing the throat thickness and making the joint unacceptable (Figure 8.8). Where fillet weld is specified at the rounded toe of a rolled section, the maximum size of the weld should generally be limited to three-fourth of the thickness of the section at the toe (Figure 8.9).

In the case of end fillet weld, which is applied normal to the direction of the force, the weld legs should be of unequal size (chamfered) with the throat thickness of the weld not less than 0.5*t*, where *t* is the thickness of the part. The chamfer in the weld is to be made in a uniform slope. A typical detail is shown in Figure 8.10.

8.3.3 Effective Throat Thickness

Throat thickness is the key dimension for calculating the strength of a fillet weld. It is defined as the height of the triangle inscribed within the weld with the apex of the triangle at the root and measured perpendicular to its outer side.

Typically, there are three face profiles of a fillet weld, namely, the miter fillet (flat face), the convex fillet, and the concave fillet (Figure 8.11). Since the capacity of a fillet weld is based on throat thickness, it is obvious that depositing too much weld metal only makes the profile of the weld to be too

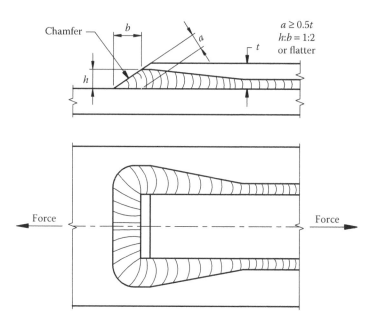

FIGURE 8.10
End fillet weld normal to direction of force.

convex, without adding any strength and also rendering the connection to be wasteful. Similarly, depositing too little metal makes the profile too concave, resulting in an inadequate joint. The correct profile should leave the weld slightly convex.

Most fillet welds are made with fusion faces at right angles (90°). However, this angle may vary from 60° to 120°. Consequently, the throat thickness will also vary with the angle between the fusion faces. Effective throat thicknesses of various fillet weld profiles are illustrated in Figure 8.11. Table 8.1 shows the throat thickness factors for a convex or miter (flat face) fillet weld of equal leg length for different angles between the fusion faces. These factors are to be multiplied with the leg lengths of welds to derive the corresponding throat thickness. Welds with unequal leg lengths are uncommon, but are theoretically acceptable. In such cases, the throat thickness should be appropriately calculated, which should not exceed 0.7 multiplied by the shorter of the two legs (Figure 8.12).

In cases where the angle between the fusion faces is less than 60° or more than 120°, fillet weld is not recommended, because of poor access in the case of the former and too small throat thickness in the case of the latter. However, in case it becomes absolutely necessary to provide such a fillet weld, appropriate tests must be performed to prove the strength of the particular fillet weld.

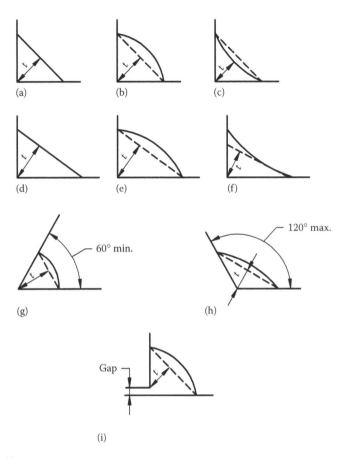

FIGURE 8.11
Effective throat thicknesses of different fillet weld profiles: (a) miter (equal legs), (b) convex (equal legs), (c) concave (equal legs), (d) miter (unequal legs), (e) convex (unequal legs), (f) concave (unequal legs), (g) acute angle convex (equal legs), (h) obtuse angle convex (equal legs), and (i) plate gap (convex)—where t is the throat thickness.

TABLE 8.1

Effective Throat Thickness Factors

Angle between Fusion Faces	Throat Thickness Factors as Multiple of Leg Length
60°–90°	0.70
91°–100°	0.65
101°–106°	0.60
107°–113°	0.55
114°–120°	0.50

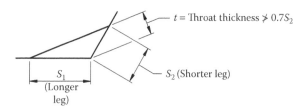

FIGURE 8.12
Throat thickness for unequal legs of fillet weld.

For concave fillets, the minimum throat thickness and the minimum leg length should be specified in the design and drawing.

8.3.4 Effective Length

In a fillet weld, there is a possibility of poor penetration at the beginning and formation of craters at the end of the weld, thereby reducing the theoretical cross section of the weld at these locations. In order to counter such a possibility, it is a common practice to reduce the actual length of the weld run by twice the weld size (i.e., one weld size for each end) to arrive at the effective length for design purpose. Also, the effective length of any individual fillet weld should not be less than four times the weld size. In case it is not possible to meet this requirement, the size of the weld should be reduced suitably and the strength of the joint checked accordingly, or an alternative detail adopted to comply with the requirement.

8.3.5 Strength of Fillet Weld

In a fillet weld, internal stress distribution pattern is rather complex due to various factors, namely, presence of multiaxial stresses, shear lag, variation in yield stress, residual stresses, stress concentration, and hardening effects. In practice, however, these complex situations are neglected and a simplified concept of transmission of only the external forces by the fillet weld is considered. It is further assumed that the weld material is homogeneous, isotropic, and elastic, and the connecting components are rigid and free form any deformation. Thus, for the purpose of computing the strength of the weld, forces are assumed to be transmitted by uniformly distributed shear stress through the weld throat area, which is defined as the throat thickness multiplied by the effective length of the fillet weld (see Figure 8.5).

Strength of the fillet welded joint is based on the strength of the steel of the connecting components and the shear strength characteristics of the matching weld metal. Normally, the weld metal provided is stronger than the connecting parent metal.

8.3.6 Design Procedure

For design of the fillet weld, the following step-by-step procedure may be adopted:

- Assume a size of weld, considering the thickness of the connecting members.
- Calculate the design strength of the weld per unit length considering the effective area of the fillet weld.
- Calculate the effective length of the weld required to be provided by equating the design strength of the weld to the external factored load.
- Prepare a detail showing the disposition of the calculated weld length in the joint.

The various aspects that need to be considered for design and detailing of the fillet welded joint are discussed in the following paragraphs.

8.3.7 End Return

Wherever practicable, fillet weld terminating at ends should be returned continuously around sharp corners or edges in the same plane for a distance of at least twice the leg length. This detail would relieve the high stress concentration at ends of weld and is particularly important on the tension end parts of members carrying bending loads, for example, brackets, beam seating, and similar parts. In such a case, the reduction of the theoretical length due to possible craters is not necessary for the end where end return has been provided (see Figure 8.13).

8.3.8 Lap Joint in End Connection

When welded lap joint is used to transfer forces, the minimum overlap should not be less than four times the thickness of the thinner part joined or 40 mm, whichever is more. In that case, the level of induced eccentricities is not likely to create appreciable secondary stresses in the weld and may be ignored.

Single-fillet weld should not be used where lapped parts are not restrained from opening up or separating. Overlapping parts should be connected by at least two lines of longitudinal (side) or transverse fillet welds.

8.3.8.1 Longitudinal Fillet Weld

Where only parallel longitudinal side fillet welds are provided, considerable variations in tensile stress occur across the width of the connecting members. A typical tensile stress distribution pattern in the members is shown in Figure 8.14. The shear stress distribution in the joint is also nonuniform, with

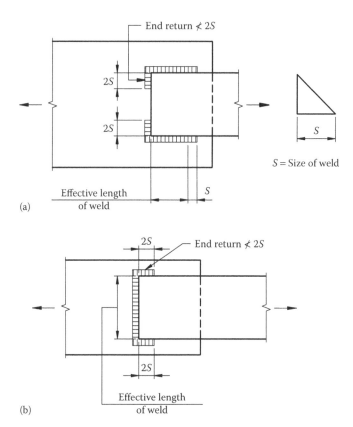

FIGURE 8.13
(a, b) End return of fillet weld.

significant concentration of stresses at the ends than at the central region. The level of the variation of the shear stress distribution depends on the length of the weld and also on the ratio of the spacing between the weld runs vis-à-vis the weld length. The nonuniformity of the stress distribution increases as the perpendicular distance between the weld run increases. In order to avoid development of serious stress concentration at the ends, the length of each side fillet weld should not be less than the perpendicular distance between them. (i.e., the width of the connected plate). For the same reason, also this perpendicular distance between the longitudinal welds should not exceed 16 times the thickness of the thinner part connected (see Figure 8.15). Where this criterion cannot be achieved, intermediate slot welds should be provided to prevent buckling or separation of the two parts.

In long side fillet welds, as in long riveted or bolted joints, the differential strain causes the shear stresses along the throat plane at the ends of the weld run to be significantly higher than those in the middle. However, they tend

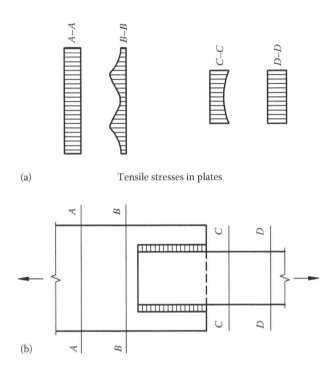

(a)   Tensile stresses in plates

(b)

FIGURE 8.14
(a, b) Tensile stress distribution pattern in plates with side fillet welds.

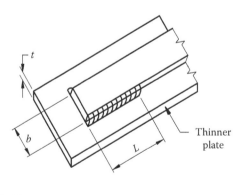

$L \not< b$
$b \not> 16t$

FIGURE 8.15
Longitudinal side fillet welds, where L is the length, b is the perpendicular distance between side fillet welds, and t is the thickness of the thinner plate.

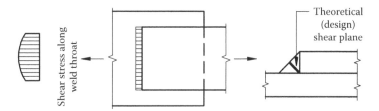

FIGURE 8.16
Stress distribution pattern in transverse fillet weld.

to become more uniform as the ultimate load is approached. Some authorities disregard this stress variation in normal welded structures, while some impose a reduction in average shear strength, if the maximum length of the side welds transferring shear along its length exceeds 150 times the throat size of the weld. The designer should go by the recommendations of the code being followed in the design.

8.3.8.2 Transverse Fillet Weld

Where only transverse fillet welds are used in an end connection, both the ends of the components should be welded. Where the distance between the transverse fillet welds is more than 16 times the thickness of the thinner part, intermediate slot welds should be used to preclude separation or buckling of the two parts.

A typical stress distribution pattern for transverse fillet welds in a lap joint is shown in Figure 8.16.

8.3.9 Combined Stresses in Fillet Weld

Often fillet welds are subjected to oblique or multidirectional loadings, leading to complex stress conditions. Examples are diagonal fillet weld in a lap joint or tee or corner joints having both transverse and longitudinal loads. In such situations, vectorial summation of stresses is required to be done to determine the stress for which fillet welds should be designed.

8.3.10 Packing in Fillet Welded Joint

In case of packing between the two components being joined by fillet weld, the following guidelines are recommended:

1. Where the thickness of the packing is less than 6 mm thick or is too thin to allow for provision for adequate welds necessary to transmit the force, the required leg length is to be increased by the thickness

of the packing. Also, the packing needs to be trimmed flush with the edges of the components to be welded.

2. Where the thickness of the packing is equal to or more than the size of the weld necessary to transmit the force, the packing should be extended and both the components should be connected to the packing individually by means of welds capable of transmitting the design force.

8.3.11 Bending about a Single Fillet

When a single-sided fillet weld is used, the joint must not be subjected to bending moment about the longitudinal axis of the weld, as this would cause the root of the weld to be in tension (Figure 8.17).

8.3.12 Fillet Weld in Compression

Figure 8.18 illustrates a joint subjected to compression forces. While designing the fillet welds, the total compression force should be considered to be transmitted by the fillet welds, ignoring the bearing contact between the components, unless provisions are made to ensure this. If necessary, full penetration weld should be considered.

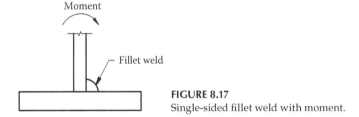

FIGURE 8.17
Single-sided fillet weld with moment.

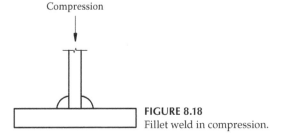

FIGURE 8.18
Fillet weld in compression.

8.3.13 Intermittent Fillet Weld

Providing intermittent fillet welds instead of a continuous fillet weld has two main advantages: saving cost and reducing the amount of distortion of the parent metal. This type of fillet weld is economical when adopted in connections, where the length of the smallest acceptable size of continuous fillet weld required to transmit the force is less than the total length of the joint. Common example of intermittent welds is the connection of intermediate stiffeners to the web of plate girders. However, in locations vulnerable to moisture ingress and corrosion, this type of weld should not be used. Thus, in the case of interior of box sections, where the connection is protected from weather, intermittent fillet weld is permitted. Also, this type of fillet weld should not be used in structure subjected to frequent variations of stresses, such as in crane girders and wind girders to avoid the possibility of fatigue failure. Furthermore, intermittent welding is not suitable where automatic welding process is used.

For designing an intermittent weld scheme, the fillet weld length required to transmit the force for the size of the weld is first computed as a normal continuous fillet weld. A series of intermittent fillet welds of total effective length equivalent to the computed length is then provided.

Two types of common intermittent fillet welds are shown in Figure 8.19. Detail shown in Figure 8.19b is generally considered preferable, since this detail tends to reduce distortion because of the balancing nature of the weld disposition.

Most authorities recommend that at any section of an intermittent fillet welded joint, the fillet weld should have an effective length of not less than four times the weld size or 40 mm, whichever is greater.

The clear spacing between the effective lengths of intermittent fillet welds should not exceed 12t for compression elements and 16t for tension elements, where t is thickness of the thinner part joined. In no case should this gap be more than 200 mm.

In built-up members, in which two plates are connected by intermittent welds, the longitudinal fillet welds should be provided at the end of each

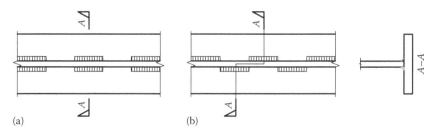

FIGURE 8.19
Intermittent fillet welds.

side of the plate for a length of at least the width of the narrower component joined. Otherwise, end transverse weld should be used, in which case the total of the end longitudinal and the end transverse welds should be at least twice the width of the narrower component. In exceptional cases, where these stipulations are not practicable, welds in slots or holes are recommended to prevent separation of the elements.

8.3.14 Analysis of Typical Fillet Welded Eccentric Connections

In this section, the basic methods of analysis for a few typical fillet welded eccentric connections, which are commonly used in design of structures are presented.

In a fillet welded joint, the weld is a continuous connection and the connecting members are assumed to be rigid.

Two types of eccentric connections are considered here:

- Load lying in the plane of the weld
- Load not lying in the plane of the weld

In both these types, the fillet welds are subjected to shear due to direct load and moment.

8.3.14.1 Load Lying in the Plane of the Weld

Figure 8.20a shows a bracket connection in which a load P is applied in the plane of the connection at an eccentricity e from the centroid G. This eccentric load is equivalent to a concentric force P passing through G and a moment $P.e$, which tends to rotate the side plate about G. For the purpose of analysis, these two load conditions may be treated separately and then the results can be superimposed to get the combined effect.

8.3.14.1.1 Concentric Force

As the connection is assumed to be rigid, and the line of force P passes through the centroid of the symmetrical weld group, the shear due to P is assumed to be uniform throughout the weld. Assuming the weld to be of unit leg length, the shear force F_a in the weld due to P is given by

$$F_a = \frac{P}{L} \tag{8.1}$$

where:
  L is the total effective length of the weld

8.3.14.1.2 Moment

The force in the weld due to the moment is considered to be directly proportional to its distance from the centroid of the weld. Thus, the weld farthest

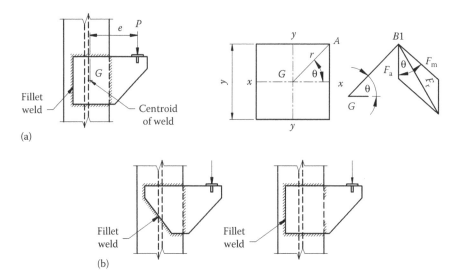

FIGURE 8.20
Eccentric welded connections (load lying in the plane of weld): (a) joints with symmetrical weld and (b) joints with asymmetrical weld.

from *G* will carry the maximum load. In this case, the weld at point *A* is farthest from *G*. The torsional moment *P·e* produces bending forces F_m in the weld about an axis passing through *G* and perpendicular to the plane of the weld, and is given by

$$F_m = \frac{P \cdot e \cdot r}{I_p} \tag{8.2}$$

where:
  r is the distance of a point under consideration from *G*
  I_p is the polar moment of inertia of the weld group

and is given by

$$I_p = I_{xx} + I_{yy} \tag{8.3}$$

$$I_{xx} = 2 \cdot \left(\frac{y^3}{12}\right) + 2 \cdot x \left(\frac{y}{2}\right)^2 \tag{8.4}$$

$$I_{yy} = 2 \cdot \left(\frac{x^3}{12}\right) + 2 \cdot y \left(\frac{x}{2}\right)^2 \tag{8.5}$$

where:

x and *y* are the lengths of the weld along *XX* and *YY* axes, respectively

In Figure 8.20a, *A* is a point farthest from the centroid of the weld and is the maximum loaded weld. The distance of *A* from *G* is given by

$$r = \sqrt{\left(\frac{x}{2}\right)^2 + \left(\frac{y}{2}\right)^2} = \frac{1}{2}\sqrt{x^2 + y^2} \qquad (8.6)$$

8.3.14.1.3 Combined Effect

The two forces at *A*, namely, F_a due to direct load and F_m due to moment have different directions. The resultant force F_r at *A* is given by

$$F_r = \sqrt{\left(F_a\right)^2 + \left(F_m\right)^2 + 2 \cdot F_a \cdot F_m \cdot \cos\theta} \qquad (8.7)$$

where:

θ is the angle between *AG* and *XX* axis

In case of unsymmetrical weld layout as shown in Figure 8.20b, the centroid of the weld is to be calculated first to determine the values of eccentricity and the polar moment of inertia. The stresses can be obtained following the same procedure.

8.3.14.2 Load Not Lying in the Plane of Welds

A commonly used bracket connection is shown in Figure 8.21. In this case also, the eccentric load *P*, applied at a distance *e* from the plane of the weld, can be considered equivalent to a direct force *P*, passing through the weld and a moment *P.e*, which tends to rotate the joint across the plane of the weld. The load on the weld can be obtained by beam bending formula.

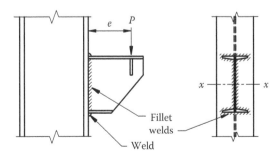

FIGURE 8.21
Eccentric welded connection (load not lying in the plane of weld).

Assuming the weld to be of unit leg length and uniform throughout, the shear force in the weld due to direct force P is given by

$$F_a = \frac{P}{\text{Effective length of weld}} \qquad (8.8)$$

The load due to moment is given by

$$F_m = \frac{P \cdot e \cdot y}{I} \qquad (8.9)$$

where:
   I is the moment of inertia of the weld
   y is the distance of the farthest weld from the neutral axis of the weld
       layout

The resultant force F_r is given by

$$F_r = \sqrt{\left(F_a\right)^2 + \left(F_m\right)^2} \qquad (8.10)$$

8.3.15 Fillet Welds in Slots or Holes

This is a method for increasing the length of fillet weld in a connection where loads are heavy and sufficient welding length is not available along the edges. The method should only be used where unavoidable, since it is somewhat difficult to lay fillet welds in slots or holes. A typical example of slot is shown in Figure 8.22.

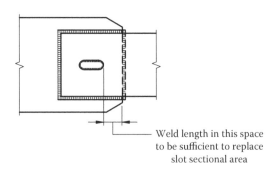

Weld length in this space
to be sufficient to replace
slot sectional area

FIGURE 8.22
Typical example of fillet weld in slot.

Recommended dimensions of the slot or hole are given here:

- The width or diameter should not be less than three times the thickness of the part in which the slot or hole is formed or 25 mm, whichever is greater.
- Corners at the ends of the slots should be rounded with a maximum radius of not less than 1.5 times the thickness of the part or 12 mm, whichever is greater.
- The distance between the edge of the part and the edge of the slot or hole or between adjacent slots or holes should not be less than twice the plate thickness or 25 mm, whichever is greater.

The first two items mentioned earlier would ensure that welders can deposit weld metal around the perimeter without difficulty. As regards the third item, it should also be ensured that there is sufficient length of weld metal in advance of the slot to transmit load equivalent to the loss of sectional area of the plate due to the slot.

Longitudinal fillet welds in the slots are considered to have the same strength as other normal fillet welds in the joint. Strength of the joint is therefore the sum of the individual capacities of all these welds.

Slots and holes should not be completely filled with weld metal or partially filled in such a manner as to form a direct weld metal connection between opposite walls, except that the welds along opposite walls may overlap each other for a distance of one-fourth of their length.

8.4 Concluding Remarks

The basic difference between a butt welded joint and a fillet welded joint is that in the former, the joint is made within the surface profile of the component members, while in the latter case, the joint is done by depositing weld metal outside the profile of the joining components. Consequently, in the design of a butt weld, there is hardly any calculation to be made. Choice of edge preparation and matching electrodes vis-à-vis, the thickness and properties of the joining components warrant primary attention of the designer and the fabricator. In the case of fillet weld, however, computation of the design strength, as well as detailing the joint layout showing the disposition of the calculated weld length is necessary at the design stage itself. Although in the drawing, fillet weld is shown as the size of the leg of the fillet, the designer uses the throat thickness of the fillet as the criterion of strength. This throat thickness is proportional to the angle between the fusion faces. The designer has to consider this aspect also. The criteria for design of fillet welds subjected to tension, compression, and multidirectional loading need to be considered in the design.

Bibliography

1. Blodgett, O.W., 2002, *Design of Welded Structures*, The James F. Lincoln Arc Welding Foundation, Cleveland, OH.
2. Dowling, P.J., Knowles, P.R., and Owen, G.W. (eds.), 1988, *Structural Steel Design*, The Steel Construction Institute, London.
3. Englekirk, R., 1994, *Steel Structures Controlling Behaviour through Design*, John Wiley & Sons, New York.
4. Hicks, J., 2001, *Welded Design—Theory and Practice*, Abington Publishing, London.
5. Hicks, J., 1999, *Welded Joint Design*, 3rd Edition, Industrial Press, New York.
6. Mukhanov, K.K., 1968, *Design of Metal Structures*, MIR Publishers, Moscow, Russia.
7. Ghosh, U.K., 2006, *Design and Construction of Steel Bridges*, Taylor & Francis Group, London.
8. Subramanian, N., 2008, *Design of Steel Structures*, Oxford University Press, New Delhi, India.

9

Fatigue in Welded Joints

ABSTRACT   The chapter begins with an introduction of the mechanism of fatigue and its effects on welded joints (fatigue cracks). Implications of fatigue on the design of welded joints are discussed, followed by deliberation on typical forms of cyclic stress loadings, stress range, stress amplitude, and stress ratio. The concept of S-N curve and step-by-step calculation of fatigue life of different types of welded joints are described. Discussion on environmental effects on structures subjected to fatigue loading is also included, followed by measures to prevent fatigue cracks. The chapter ends with a short description of the methods for improvement of fatigue performance of already-fabricated welded joints.

9.1 Introduction

Many welded structures operate under fluctuating stresses. Common examples of such structures are as follows:

Structures	Situations Causing Fluctuating Stresses
1. Cranes and crane girders	Lifting and lowering of loads, and moving with loads
2. Bridges	Heavy moving loads such as loaded truck and goods train
3. Offshore structures	Wave movement
4. Lock gates	Tidal variations
5. Towers/pylons	Wind flow
6. Screens for grading gravel and coal	Vibration

Fluctuations may be slow or fast. Examples of the former are stresses on a lock gate in a port and on a crane girder in an industrial building. Vibrations caused on screens for grading gravel and coal constitute fast type of fluctuating stresses. Fluctuating stresses may also be induced by resonance at the natural frequency of a structure. Fluctuating stresses, slow as well as fast types, reduce the ultimate strength of the material considerably, resulting in failure (crack or fracture) occurring at stress levels much lower than the yield stress. This phenomenon of localized permanent structural change in a material subjected to conditions, which produce fluctuating stresses that may initiate and/or propagate cracks in a member after sufficient number of fluctuations, is termed *fatigue*. Thus, a member, which may withstand a single application of a particular load, may fail if the same or even a smaller load is repeated for a large number of times, say 10 million times.

9.2 Fatigue Crack

9.2.1 Causes of Fatigue Crack

Welded joints are particularly susceptible to fatigue cracks. The main reasons are discussed here.

9.2.1.1 Stress Concentration

In a welded joint, sharp changes in the direction of the profile generally occur at the toes of butt welds and at the toes and roots of fillet welds. These locations are the points of stress concentration. Small discontinuities such as *undercuts* (Chapter 3) at these locations lead to high local stresses and consequently grow faster, initiating cracks (Figure 9.1).

9.2.1.2 Intrusions

Close microscopic examination at heat-affected zone and the solidified weld metal (Chapter 2) has revealed that severe heating and cooling cycle during welding leaves at the toe of the weld small irregularities, in the shape of crack-like features or tiny surface cavities filled with slags. These are termed *intrusions*, and are typically 0.15 mm deep (Figure 9.2). These intrusions are not as deep as the undercuts, and are normally not detected by conventional nondestructive test (NDT). However, these are capable of acting as the focal points for triggering fatigue cracks in a welded structure.

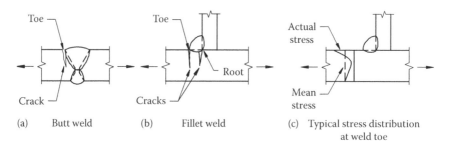

(a)   Butt weld   (b)   Fillet weld   (c)   Typical stress distribution
                                                              at weld toe

FIGURE 9.1
(a–c) Local stress concentrations at welds.

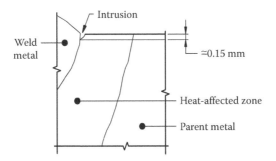

FIGURE 9.2
Weld toe detail showing intrusion.

9.2.2 Crack Growth Rate

Initially, fatigue cracks generally remain very small for a long period of time, and suddenly grow very quickly so as to be visible by naked eyes. However, even then, these may not be apparent to untrained eyes, until grease is accumulated or rust is formed in the crack. Thus, when a crack is detected, a considerable part of the life of the structure may have already been consumed, and a significant portion of the cross-sectional area of the affected component may have been lost. Failure occurs usually in the tension region when the remaining area is no longer sufficient to carry the applied load. This failure load would be well below the value calculated for static loading condition. This unwelcome situation presents considerable problems for routine inspection of such structures.

One other factor needs mention here. In case of a welded joint, a crack developed at a weld tends to progress and may affect both the connecting components, thereby damaging the entire structure, unlike in a riveted or bolted connection where a crack normally stops at the hole, and in most cases does not travel beyond to the connecting component. Thus fatigue cracks in welded structures call for special attention, unlike in riveted or bolted structures.

9.3 Design

9.3.1 Implications on Design

Unlike static design, design for fatigue in welded joints involves certain practical implications. As for example, design for fatigue is dependent on the loading sequence in the entire service life of a structure, rather than only one critical load condition. Thus, design for fatigue requires fairly accurate prediction of loads and consequent elastic stresses. Also, geometrical details rather than mechanical properties play an important role in the design of satisfactory welded joints. In fact, the geometry and the resulting stress concentration effects are the most important aspects that influence the fatigue resistance of any welded joint. Furthermore, superior workmanship and close inspection during the welded fabrication process are of utmost importance in welded steelwork subjected to fatigue loading during service life. Some of the causes that may trigger fatigue failure due to defects in welding are porosity and slag inclusions, lack of proper fusion or microscopic cracks, and crystalline change in the heat-affected zone in the base metal. It is therefore necessary that such structures be critically examined at the design stage itself and adequately provided for in the design. This critical examination must cover not only the main structural connections but also all welded attachments, even though some of these might appear insignificant.

9.3.2 Design Method

Figure 9.3 illustrates the typical forms of cyclic stress loadings in a member. The figure shows stress pattern for a member subjected to variation in tensile stresses and reversal of stresses (tension and compression). As is shown in these figures, *stress range* is the difference between the maximum and minimum stresses in the cycle, and *stress amplitude* is half the stress range. The term *stress ratio* is defined as the ratio between *minimum stress in a cycle* and *maximum stress in the corresponding cycle*.

Fatigue strength is normally presented in the form of curves of stress range against cycles to failure, commonly known as *S-N curves*. A typical S-N curve is shown in Figure 9.4, in which the total cycle stress range (S) is plotted against the number of cycles (N). S-N curves are based on test data obtained experimentally in the laboratory by conducting tests on a series of identical test pieces of different types (categories/classes) of welded joints subjected to cycles of constant amplitude and measuring the number of cycles required for failure. The category (class) of the detail depends not only on the type and detail of the weld adopted in the particular joint but also on the direction of the stress vis-à-vis the weld. For example, a butt weld detail may be same for two joints, but the categories of the S-N curves may be different depending on the direction of the stress and the location of the potential crack. Failure occurs when the remaining cross section in the specimen is inadequate to support the load. The number of cycles that causes such failure is termed *fatigue life* of the joint. It may be observed from the S-N curve that increasing the stress range reduces the fatigue life. The results of the tests are plotted using logarithmic scales for both axes because log S-log N relationship for many materials is approximately linear.

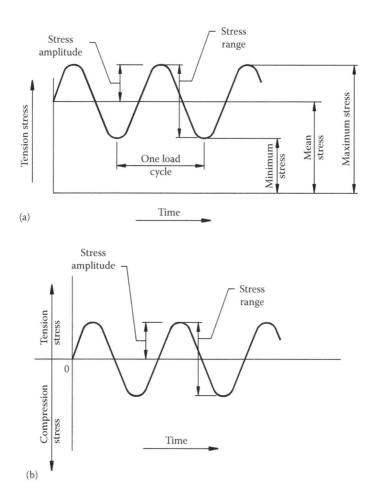

FIGURE 9.3
Typical cyclic stress loadings: (a) fluctuating tension (mean stress > 0) and (b) fluctuating tension compression (mean stress = 0).

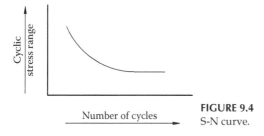

FIGURE 9.4
S-N curve.

Summarizing, in designing members and welded joints, the factors to be considered are as follows:

- Number of loading cycles and stress range
- Category or class of the joint depending on constructional details, location of the joint, and direction of stress

Calculation of fatigue life involves a few simple steps. First, it will be necessary to obtain stress history on the detail. The next step is to identify the particular S-N curve for the particular weld detail. Having identified the appropriate S-N curve for the particular weld detail, the stress range is located in the vertical axis and reading across the curve, the corresponding fatigue life (in number of cycles) is found on the horizontal axis of the curve.

Fatigue design rules used by different countries are based on the results of laboratory tests carried out in specimens of different weld configurations corresponding to different combinations of stress range and cycle loadings. These are to be followed while designing members and joints to cater for fatigue effects in welded structures.

In the preceding paragraphs, loading has been considered as a single fluctuating load of constant amplitude. In actual practice, however, structures are often subjected to more than one type of loading and each type may vary in intensity. In such cases of loading producing nonuniform stress range cycle, the sequence is broken into a simplified stress range spectrum consisting of bands of constant stress level by a rational method and the fatigue life is obtained by following the procedure specified in the particular code being followed in the design.

9.4 Environmental Effects

Studies reveal that the rate of growth of fatigue cracks depends on local environment also. As for example, where structural components are exposed to corrosion-prone humid environment, the synergic effect of corrosion and fatigue, known as *corrosion fatigue*, becomes a serious problem. In such cases, the rate of growth of a crack will be more than that in a dry environment or in the vacuum.

Also, the rate of crack development in marine and offshore structures will be different. Tests have shown that presence of seawater, which is a complex mixture of substances, has significant effect on the growth of crack in steel. Some authorities recognize this phenomenon and have incorporated the same in their recommendations. The use of cathodic corrosion protection system in marine and offshore structures can help to impede the crack growth rate in such structures Thus, adequate attention needs to be given to the effects of environment of welded structures. Fatigue data for the specific environment should be used in order to reduce the chances of fatigue failure.

9.5 Prevention of Fatigue Cracks

To avoid fatigue cracks, the factors that are of greatest importance are the design and details of the welded joints. Some of the points that need to be addressed are briefly discussed here.

In structures subjected to cyclic loading, fatigue failure occurs in the tension zone of members and is generally initiated by stress concentration. Some examples of such stress raisers are as follows:

- Locations with sharp changes in cross sections, sharp corners, notches, and re-entrant angles
- Rough flame-cut edges
- Excessive or inadequate reinforcement
- Arc strikes
- Accidental dents
- Welded attachment in highly stressed zones

  Note: Where welded attachments in highly stressed zones are unavoidable, the weld profile should be dressed to merge with the parent metal.

While designing welded joints, special attention should be paid to the following aspects:

1. Fillet welds should not be located at right angles to the line of principal stresses in tension members.
2. Intersections of longitudinal and transverse welds should be avoided.
3. Details that produce severe stress concentration or poor stress distribution should be avoided. Thus, gradual transitions in sections should be adopted in preference to abrupt changes in sections (see item 15 below). Also, weld details, which induce high localized constraints, should be avoided.
4. Re-entrant notch-like corner details should be avoided.
5. Butt welds are preferable to fillet welds.

6. Edge preparation for butt welding should be so detailed that minimum weld metal is required. This, in turn, would minimize locked in stress in the joint.

7. Size of fillet welds should not be larger than is required from design consideration.

8. Deep penetration fillet weld is preferable to normal fillet weld.

9. Except for connecting intermediate stiffeners to webs of beams and plate girders, intermittent fillet welds should not be used in structures subjected to fatigue conditions.

10. Eccentricities in welded connection details should be kept to the minimum, and if possible, should be avoided altogether.

11. In structures subjected to fatigue loading, it is recommended to provide multiple load paths (redundancy) to avoid overall collapse of the structure in case one element fails due to fatigue.

12. Unsealed butt welds of single-V, U, J, and bevel types and incomplete penetration butt welds should not be used. A sealing run of weld metal should be deposited on the backs of single-V, U, J, bevel, or square butt welds. In cases where such sealing runs are not practicable, steel backup strips should be provided and welded to the joining components. Typical single-V butt weld with backing strip is shown in Figure 9.5.

13. Intermittent butt welds should not be used.

14. In multi-run butt welds, which are to be welded from both sides, the back of the first run should be gouged out by suitable means before welding is started on the gouged out side.

15. *Butt welding parts of unequal sections:* When two plates of different thicknesses are required to be butt welded, if the difference in thickness of the plates exceeds 25% of the thickness of the thinner plate, or 3 mm, whichever is greater, the thicker plate should be reduced to the thickness of the thinner plate outside the butt weld detail by providing a slope of not steeper than 1 in 5. See Chapter 11, Figure 11.2. In case the difference in thickness is less than 25% of the thickness of the thinner plate or 3 mm, whichever is greater, the transition in thickness may be accomplished by sloping the

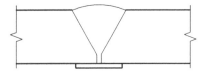

FIGURE 9.5
Single-V butt weld with backing strip.

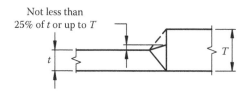

Not less than
25% of *t* or up to *T*

FIGURE 9.6
Butt weld detail where transition is not possible. *T* is the thickness of the thicker plate; *t* is the thickness of the thinner plate.

weld faces by chamfering or by combination of the two methods at an angle not steeper than 1 in 5. Where reduction of the thickness is impracticable, the weld metal should be built up at the joint to not less than 25% of the thickness of the thinner plate or up to the dimension of the thicker plate. This has been illustrated in Figure 9.6.

16. *Butt welded* T *joints:* Butt weld in T joints (typically in a plate girder web to flange connection) should be reinforced by welding with a minimum of 25% of the thickness of the outstanding part, but not greater than 10 mm, as shown in Figure 9.7.

17. *Ends of butt welds:* In order to ensure full throat thickness at the beginning and end of butt welds, it is a common practice to provide run-on and run-off plates of the same thickness of the parts joined at both ends of the butt weld, past the edges of the parts joined. These are typically of lengths of not less than 40 mm. These run-on and run-off plates are generally removed by abrasive cut off or by hacksaw blade, after the welding is completed, and the ends of the weld finished smooth and flush with the edges of the abutting parts. Oxy-acetylene cut should not be encouraged in order to avoid possible thermal stresses. However, in case flame cutting is used, cutting should be at least 3 mm from the parent metal, and the remaining metal removed by grinding or any other approved method.

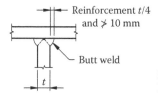

Reinforcement *t*/4
and ⊁ 10 mm

Butt weld

FIGURE 9.7
Butt welded *T* joint.

18. *Reinforcement of butt weld:* In order to ensure adequate throat thickness is available in a butt weld, it is a practice to provide sufficient reinforcement, but not exceeding 3 mm to a butt weld. However, where a flush surface is required, care should be taken so as not to reduce the thickness of the butt weld at any location.

9.6 Improvement of Welded Joints

Fatigue performance of already-made welded joints can be enhanced by

- Reducing stress concentration in the weld
- Removing crack-like defects in the toe of the weld
- Reducing residual tensile stress
- Introducing compressive stress

There are a number of methods for achieving one or more of these situations. Some of these are discussed here.

9.6.1 Grinding

Unwelcome presence of intrusions near the toe of some welds has been discussed earlier. Some authorities suggest that these intrusions may be removed by grinding. This would reduce the chance of initiating fatigue cracks. However, in doing so, care should be taken to ensure that the grinding is deep enough to remove the intrusions, and at the same time, it does not bring on the surface of the weld internal weld defects such as porosity, as it might again make the joint liable to possible fatigue crack. Also, grinding should be done carefully so as not to impair the intended strength of the weld.

9.6.2 Peening

Another method to improve the fatigue life of a welded joint is to introduce residual compressive stresses in the surface around the weld toe, where fatigue cracks are likely to initiate. The method is called *peening*. The area around the weld toe is *hammered* with a round-headed peening hammer or needle gun. Alternatively, shot blasting or ultrasonic impact peening treatment may also be done to obtain similar effect. While introducing residual compressive stress, peening also improves the fatigue properties by flattening crack-like defects at the toe and providing improved toe profile. However, as with any other method, the treatment may prove to be counterproductive and damaging, if not done properly.

9.6.3 Dressing

Another effective treatment for improving the fatigue life of a welded joint is to dress the toe with a tungsten inert gas (TIG) or plasma TIG torch by melting out the intrusions and providing a smooth blend between the weld face and the parent metal. This method also removes slag inclusions and undercuts at the toe. However, this is a rather slow process and suitable for smaller work only.

It should be borne in mind that the methods described earlier need considerable supervision and quality control effort to be successful. Otherwise, the results may negate the very purpose of the operation. It is, therefore, always preferable to produce a well-designed and properly executed joint, rather than produce a bad job and then try to improve it.

9.7 Concluding Remarks

In recent times, use of powerful software packages, which allow detailed finite element analysis, have enabled designers to approach the problems posed by complex fatigue loading with far more confidence. However, in spite of such advancements, the designer has still to recognize the harmful effects of fatigue in welded joints, and consider this aspect seriously for all such joints, including the non load carrying attachments, which appear to be quite unimportant for service performance.

Bibliography

1. Hicks, J., 2001, *Welded Design—Theory and Practice*, Abington Publishing, England.
2. Hicks, J., 1999, *Welded Joint Design,* 3rd Edition, Industrial Press, New York.
3. Dowling, P.J., Knowles, P.R., and Owens, G.W. (eds.), 1988, *Structural Steel Design*, The Steel Construction Institute, London.
4. British Standards BS: 5400, 1980, Part 10, Code of practice for fatigue, British Standards Institution, London.
5. Ghosh, U.K., 2006, *Design and Construction of Steel Bridges*, Taylor & Francis Group, London.

10

Beams and Columns

ABSTRACT Typical beam and column sections, both rolled and built up by welding, are introduced. The principal types of different welded connections are discussed, namely, splices in beams and columns, column base, column cap, beam-to-beam connections, and beam-to-column connections. The chapter concludes with a note on the principles of formation of different types of castellated beams and their weld details.

10.1 Introduction

This chapter covers welded connection details of beams and columns involving rolled as well as welded compound sections commonly used in construction.

10.2 Beams

10.2.1 Beam Sections

Figure 10.1 shows some typical beam sections used in light structures.

Rolled angles are commonly used for roof purlins and sheeting rails, where only light loads are to be carried. Rolled steel joists and universal beams are the most frequently used beam sections and owe their popularity to their symmetrical configuration. Welded beams built up from plates have an advantage of concentrating the material as far away as possible from the neutral axis, thereby achieving the maximum moment of inertia of the section. At the same time, in oblique bending, a channel shape behaves somewhat better than a symmetrical *I*-shaped section.

10.2.2 Splices in Beams

Members most often used in beams are rolled steel joists and universal beams. In order to make up lengths, it is necessary to introduce splices either at the shops or at site. In both cases, full penetration butt weld is preferred. For the design of splices, it is generally assumed that the applied moment is

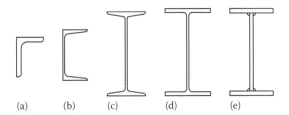

FIGURE 10.1
Types of beam sections: (a) rolled angle, (b) rolled channel, (c) rolled joist, (d) universal beam, and (e) welded beam.

carried by the flanges, and the transverse shear is transmitted by the web. For transmitting the forces through butt welds properly, weld preparations need to be made with care. However, in rolled beams, particularly in those with tapered flange profile, accurate preparation at the junctions of web and flanges is difficult to achieve. Inaccuracy in the preparation at such locations may result in a weld with incomplete penetration and consequently an unacceptable joint. In order to avoid such a situation, often a square butt weld is preferred (see Figure 10.2a).

It is a normal practice to locate the beam splice away from the area of maximum bending moment. However, in case it becomes obligatory to locate a square butt welded joint at the point of high bending moment, some authorities recommend strengthening it by providing tapered flange cover plates as shown in Figure 10.2b. The taper profile of the cover plates will facilitate

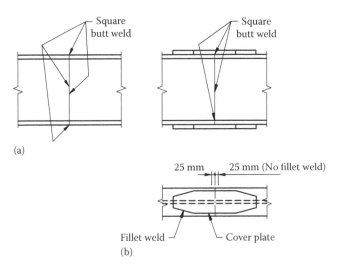

FIGURE 10.2
(a, b) Welded splice details of rolled beams.

uniform transmission of forces across the joint, thereby avoiding concentration of stress. Most codes recommend that the area of the cover plates at the joint location should be at least 5% in excess of the area of the flange elements. Also, cover plates should be so proportioned that their combined centroidal axis coincides with that of the members to avoid eccentricity. Some authorities recommend not to weld up to the center line of the joint, but to leave a gap of at least 25 mm on each side of the joint without weld, so as to avoid concentration of stress at the joint location.

Another alternative detail for splice in rolled beams is to provide a division plate between the square ended beam parts and connect both the beam parts to the division plate by fillet or butt weld all round the profile of the members. This detail is illustrated in Figure 10.3a. Needless to add the division plate should be of the same grade of steel as that of the beam parts. Also, the thickness should be at least that of the thicker flange of the beam parts. It is important to check the division plate for lamination prior to welding, as well as after welding. A variation of the detail is shown in Figure 10.3b, in which the division plate is stopped short of the tension flange, thereby providing a square slot in which butt weld can be made. This is done to keep the bottom surface flush. In this detail, it is important to ensure root penetration at the junction of the division plate and the tension flange. Otherwise, this variation should be avoided at locations of high bending moments.

For splices in built-up welded beams, arrangement of splice locations, utility of cope holes, and sequence of welding, please refer to Chapter 11 on plate girders, where these topics have been discussed in detail.

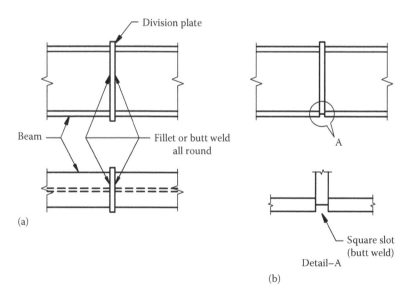

FIGURE 10.3
(a, b) Beam splice details with division plates.

10.3 Columns

10.3.1 Column Sections

Column section may be either solid column section or open web column section. A solid column section comprises one or several rolled shapes or plates connected together by welding, producing solid external surfaces. Figure 10.4a through j shows some examples of such solid columns. An open web column section consists of two or more rolled sections connected to each other with batten plates or lacings. Figure 10.4k through m illustrates some of the commonly used open web column sections. An open web column has the advantage of increased slenderness ratio by placing the main components

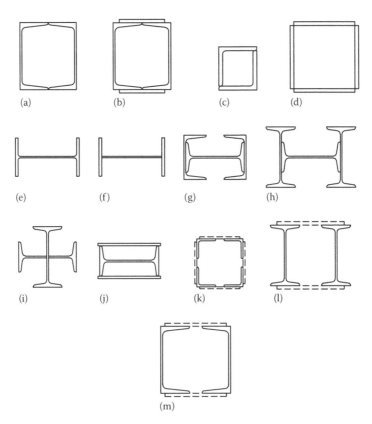

FIGURE 10.4
Types of column sections: (a) two channels, (b) two channels and two plates, (c) two angles, (d) four plates, (e) universal column, (f) three plates, (g) one beam and two channels, (h) three beams, (i) two beams, (j) universal column and two plates, (k) four angles and battens, (l) two beams and battens, and (m) two channels and battens.

away from the centroidal axes of the column section, thereby economizing on the consumption of steel material. This advantage is, however, often offset considerably by fabrication cost because of the use of short welds manually to connect the battens and/or lacings to the main component.

In column sections made up of plates only, the designer has basically two options to adopt from, namely, the *I*-section or the box section. In almost all cases, the *I*-section will prove heavier than a box section to carry a given load. However, due to the wide use of automatic welding, the fabrication cost for an *I*-section is less, although the amount of weld is somewhat the same. In long columns where radii of gyration in both the principal directions are important, the *I*-section performs rather poorly and adoption of the box section is the ideal solution. In such a case, although the fabrication costs are more than those for a simple *I*-section, the saving in the weight normally results in an overall cheaper job.

In all box sections, the insides are inaccessible after fabrication and must therefore be sealed off hermetically in order to prevent the possibility of corrosion. Consequently, in such cases, all longitudinal welds must be continuous and the ends of the components welded to cap plates, base plates, or diaphragms as the case may be. In the case of box columns made up of rolled sections, it must be ensured that both the component parts come from the same rolling batch to ensure proper fit up. Also, some additional shop work may be necessary to bring the toes of the members into proper alignment and maintain the same throughout the welding operation.

10.3.2 Eccentrically Loaded Columns

A column forms one of the basic types of load carrying component in a structure and transmits load applied at the top (cap) to the supporting structure below through the footing. In many cases, however, the column is not subjected to compression only, but is required to resist some degree of bending as well. For example, if the load at the top is not applied along the axis of the column, the column will be required to withstand not only the compression induced by the longitudinal force but also the bending moment due to eccentricity. In the corner column in a building, the column is subjected to bending about both the axes and the column is to be designed accordingly. In an industrial building, it is common to have eccentrically loaded columns supporting roof load, as well as crane load.

10.3.3 Column Weld Details

For a column carrying only axial load, the longitudinal welds connecting the individual sections together, whether box or single *I* are not generally subjected to calculable forces. The minimum weld requirement of the guiding code of practice for the related thickness of the part welded should be adequate to cover any shearing force. However, some authorities recommend

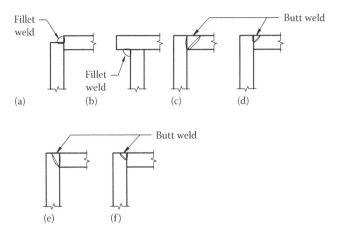

FIGURE 10.5
(a–f) Typical welding details for box columns.

that these longitudinal welds be designed for 2.5% of the axial load carried by the column unit. When a built-up column section is required to resist bending moment, due attention should be given to determine the size of the longitudinal weld connecting the component parts. For example, in the case of a column of the *I*-section, built up from plates, the welds that join the web with the flanges are subject to longitudinal shear forces resulting from the moment along the length of the column. These longitudinal welds will have to be designed to resist these shear forces.

A box column fabricated from four plates may be joined at the corners in a variety of ways. Some of these are shown in Figure 10.5. In a flush-type box construction, the weld is deposited in a bevel preparation at the edges of the two plates as shown in Figure 10.5c through f. However, to avoid chances of lamellar tearing in the connecting plate, the details shown in Figure 10.5c and d should be avoided. Cracks due to lamellar tearing have been discussed in Chapter 3.

For column sections formed by two channels, joists, or universal beams, welded toe to toe, a small bevel of about 3 mm may be made on both the connecting units at the contact region of the flanges and weld metal deposited in the bevel preparation.

The welds described in the preceding paragraphs of this section are suitable for machine welding, being long straight line of continuous weld.

10.3.4 Column Splices

Both shop and site splices are commonly used to make up lengths of columns. Shop splices are provided to make up limited lengths of available rolled sections. Site splices are generally made because of limitations in lengths in transport facility from fabrication shop to the erection site, or due to the necessity of changing the section. Site splices may also be dictated by the consideration of handling facilities available at site, particularly for light and slender members. It is a common practice to locate column splices just above the floor level. For earthquake-prone regions, it would be expedient to locate column splices in regions of near-zero bending moments.

For the purpose of design of a column splice, the axial load is considered to be shared proportionately by the combined area of the flanges and the web. The flanges carry the applied moments, while the transverse shear is carried by the web. Welding is designed accordingly. For highly stressed members, full penetration butt welding is generally preferred. However, for members subjected to low stresses, incomplete penetration butt weld may be permitted. Column splice may also be made by providing splice plates in the flanges and the web.

In a built-up column section, it is not mandatory that shop splices of components are to be loaded at a single place. Generally, individual components are spliced to make up the length before the column is assembled to its required configuration. Figure 10.6a illustrates a typical detail of flange splices in a built-up *I*-section. The simplest option for such splices is butt welding, which can be designed to ensure proper transfer of load from one element to the other, including tension due to eccentric loading condition. Figure 10.6b shows a typical shop splice using splice plates. In this case, the flange and web splices are located at the same place. In this type of splice care should be taken to ensure that the cross-sectional areas of the splice plates satisfy the requirements of the governing code. Also, the length of the fillet welds connecting the splice plates and the column units should be sufficient to transfer the load from one unit to the other. While designing a splice with splice plates, care should be taken to ensure that the centroidal axis of the splice plates coincides with that of the column units, to avoid any eccentricity

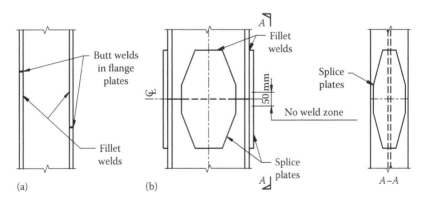

FIGURE 10.6
(a, b) Typical details of shop splices for columns.

in the joint. Some authorities recommend a *no weld zone* of about 25 mm on each side of the joint to avoid stress concentration at the column ends. Also, for the same reason (avoiding stress concentrations), ends of the splice plate are tapered as shown in Figure 10.6b. Splices with splice plates are convenient in case of rolled sections, since requisite preparation for the butt weld, particularly at the roofs is not easy to achieve and imperfect preparation may lead to inefficient joint.

Some typical site splices are shown in Figure 10.7. It is a good practice to machine both the ends of the column sections in the shop to ensure a square bearing surface at the location of site splice. Figure 10.7a shows a splice detail where both the column units are of identical cross sections. Two pairs of angle cleats are provided in the webs of the connecting column units to temporarily splice and hold the two adjacent column units together while these are being butt welded. Since these angle cleats do not extend beyond the ends of the column units, the possibility of their being damaged during transit is minimal. It should, however, be noted that in all

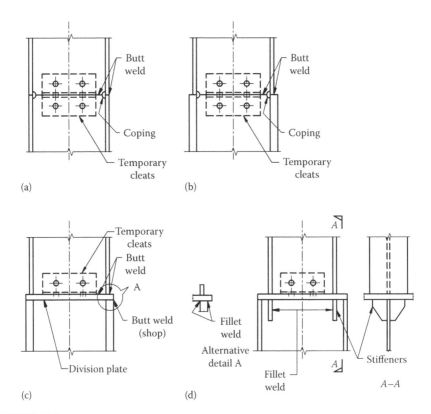

FIGURE 10.7
(a–d) Typical details of site splices for columns.

site splices, these cleats are meant to hold the two column units together in position for the purpose of welding only. For maintaining the vertical alignment of the column as a whole with respect to the rest of the structure, the upper unit must be suitably guyed or propped until the welding is completed. Figure 10.7b shows a detail where the flange of the upper column unit is thinner than that of the lower unit. Other details are similar to those as shown in Figure 10.7a. Splice detail of two dissimilar column sections with a division plate placed in between is illustrated in Figure 10.7c. In this case, the welding is continuous all round the profile of both the connecting sections. Welding may be either butt or fillet weld as shown in the figure. The division plate is normally shop welded to the lower column. Two temporary angle cleats are provided in the lower end of the upper column unit. The upper column is erected on top of the division plate and the temporary cleats are bolted to the division plate. In case the lower column unit is deeper than the upper column unit, stiffeners are normally welded in the web of the lower column directly below the flanges of the upper column. These stiffeners will provide support to the division plate, thereby reducing its thickness. Figure 10.7d shows a typical detail where such stiffeners are used. Where division plate is used, care must be taken to ensure that the plate is of the same grade as that of the column components, and its minimum thickness should be equal to that of the thicker flange. Also, the plate should be tested for lamination before and after fabrication, to ensure that it can withstand tensile force through its thickness due to any moment condition to which the columns may be subjected, both during erection and service conditions.

10.3.5 Column Bases

Column base is provided at the lower end of the column to effectively distribute the concentrated compressive load of the column over a much larger area of the concrete foundation, which supports the column. Column base is also required to ensure adequate connection of the column end to the foundation. When a column is symmetrically positioned on the base plate and carries a concentric axial load, the distribution of pressure between the base plate and the concrete foundation is uniform. However, eccentric loading on the column produces bending moment in the base, in addition to compressive force. This situation results in compression and tension zones in the contact area between the base plate and the concrete foundation. While the forces in the compression zone get distributed over the contact area, the forces in the tension zone are transferred to the concrete foundation by means of suitably proportioned holding down bolts, which are adequately anchored in the concrete. Thus, the couple formed by these two forces balances the applied axial compression and bending moment.

The horizontal shear at the base is normally transferred to the foundation either by friction at the contact surface or by holding down bolts. For transmitting

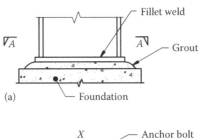

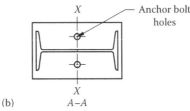

(b)   A–A

FIGURE 10.8
(a, b) Pinned column base.

high shear force, special shear key plates welded to the underside of the base plate and embedded in the concrete arc often used.

Broadly, column bases can be divided into two categories, namely, pinned and fixed types.

10.3.5.1 Pinned-Type Base

Pinned-type base system is used where there is very little tension between the base plate and the foundation, and the required restraint against angular rotation about X-X axis is minimal (i.e., pinned condition). Pinned-type bases are popular in portal frames and multistoried buildings.

A common pinned-type base for an axial loaded column is illustrated in Figure 10.8. The arrangement consists of a thick steel base plate on which the milled lower end of the column rests. The base plate is fillet welded all round the column profile and is provided with holes at the center line of the base for anchor bolts. For light columns or where end milling machines are not available, end milling is dispensed with and the entire load is transmitted to the base plate through welds. Some authorities suggest that welding all round the column profile is not necessary; welding the flanges and only the part of the web should suffice. In practice, either of the methods is acceptable, provided the strength criteria are satisfied.

10.3.5.2 Rigid-Type Base

Rigid-type (fixed-type) column base is generally used in conditions of moment at the base, and is normally detailed with a system comprising

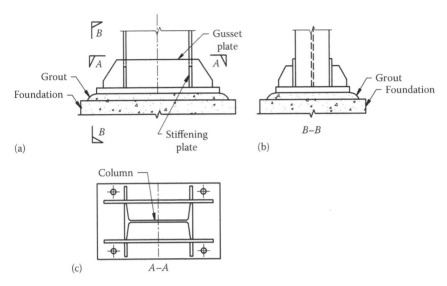

FIGURE 10.9
Pinned (a–c) Rigid column base (type I).

of welded base plate, gusset plates, and the column section. The principal function of such arrangement is to provide rigidity to the base, which prevents angular rotation of the column and allows the lever arm of the holding down bolts to be increased, thereby increasing their resistance to bending moment. This arrangement also enables the designer to keep the thickness of the base plate minimal. Fixed-type column bases are very common in industrial buildings, portal frames, and rigid framed multistoried construction. Salient features of five types of common rigid column bases are discussed here:

Type I: Figure 10.9 shows a detail where gusset plates are attached to the outer edges of the flanges of the column, and fillet welded to the column. Additional stiffening plates are attached transversely to the outer faces of the gusset plate and fillet welded to the gusset plate. This assembly is fillet welded to the base plate. Although accessibility for the welds on the inside is not so good, the base plate can be satisfactorily welded to the outer faces of the column flanges, as well as the gussets and stiffening plates. Also, the inside fillet welds between the column and the gusset plates should be done prior to assembly of the base plate.

Type II: In Figure 10.10, two channels with vertical stiffeners are attached to the outer faces of the flanges by fillet welding. For ease of fillet welding at the bottom, the channels are set back slightly from

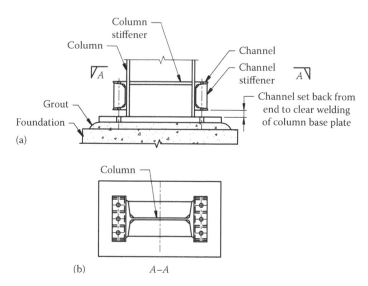

FIGURE 10.10
(a, b) Rigid column base (type II).

the machined end of the column. With this arrangement machining of the underside of the channels is not required. The holding-down bolts are located on the top flanges of the channel sections. The channels and connecting fillet welds are to be designed to transfer the vertical forces (due to moment) to the holding down bolts. Horizontal stiffener plates should be provided between the column flanges at the channel flange level to stiffen the column flange against possible local bending of the column flanges due to the resultant horizontal force.

Type III: Figure 10.11 shows a detail, which is somewhat a combination of the details shown in Figures 10.9 and 10.10. In this case, the holding down bolts are located outside the narrower base plate and the channel support system for holding down bolts is replaced by individual built-up units made with *U* cut horizontal support plates for holding down bolts placed on top of pairs of vertical stiffener plates and washer plates to be welded on the horizontal support plates after alignment of the column. Lengths of the vertical plates will depend on the length of welds required to transmit the tensile forces to the holding down bolts.

Type IV: Figure 10.12 illustrates a detail where gusset plates are fillet welded to the outer faces of the column flanges. Vertical stiffening plates are also fillet welded to both sides of the web to strengthen it.

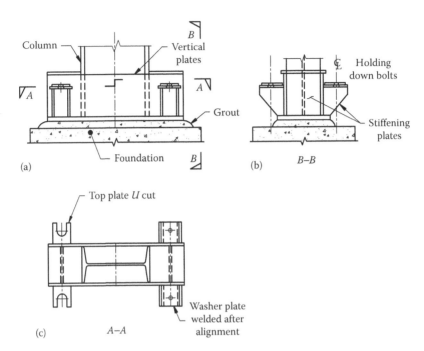

FIGURE 10.11
(a–c) Rigid column base (type III).

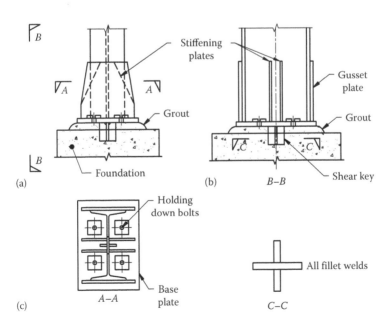

FIGURE 10.12
(a–c) Rigid column base (type IV).

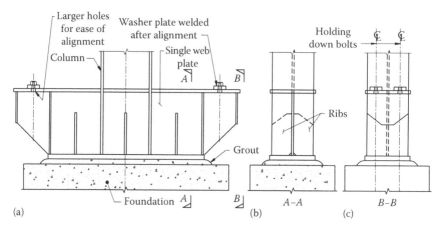

FIGURE 10.13
(a–c) Rigid column base (type V).

This assembly is then fillet welded to the base plate. Holding down bolts are located on washer plates placed on the base plate. For easy alignment, the diameters of the holes in the base plate and the washer plates are generally made 8 and 3 mm, respectively, larger than that of the holding down bolts. The washer plates are welded to the base plate after alignment of the column. The welds connecting the vertical gusset and stiffening plates to the column and the base plate are designed to transmit the forces from the column to the holding down bolts via the base plate.

Type V: In Figure 10.13, a base detail for columns with heavier load and larger moment is shown. In this case, vertical extension plates are provided at the column and the holding down bolts, located beyond the length of the base plate, are supported by horizontal plates with vertical stiffener on both sides of the column web and the vertical extension plates. The single-web base detail adopted in this case makes welding very convenient.

10.3.6 Column Caps

A flat plate fillet welded to the top end of the column is generally quite sufficient to form a column cap for transmitting load on to the column. If the load is heavy, the top end of the column may be machined before welding of the cap plate. In such cases, load on the column cap is transmitted directly by bearing and welds need only to be nominal to hold the cap with the column securely in place. If the column end is not machined, load is transmitted by the welds, which have to be designed accordingly.

10.4 Connections

10.4.1 Types of Connections

Basically, there are three methods of design and design assumptions for connections between individual frame components such as beam-to-beam and beam-to-column connections. Based on these assumptions, connections are categorized into three types, each defined by its structural behavior, namely simple, rigid, and semi-rigid.

10.4.1.1 Simple Connection

Simple connection is assumed to transfer the vertical shear reaction only, with no bending moment present at the connection. Thus, the beam behaves as a simply supported one, and will deflect under load. The connection is also detailed to allow the beam end to rotate slightly. This type of connection is often termed as a *flexible connection*, and is typically used in frames up to about five stories high, where strength rather than stiffness is the criterion for design.

10.4.1.2 Rigid Connection

Rigid connection is designed to transmit both shear and moment across the joint, which is detailed to behave as a monolithic joint. The moment is transmitted to the column itself as also to any beam located on the opposite side. Also, the connection is detailed to behave as a monolithic joint and must have sufficient rigidity to withstand rotation and maintain the original angles between connecting members.

The rigidity of a joint is dependent on the rigidity of its main support members. Thus, where beams are connected to the column flanges of comparatively thin thickness, there is a possibility of reduction of rigidity of the connection, unless it is stiffened by providing stiffeners between the column flanges in line with the beam flanges. Similarly, for a beam framing into a column web, it would be necessary to weld the flanges of the beam to the column web directly to provide a rigid connection. Alternatively, suitable plates should be provided to connect the flanges to the column web by welding. Thus, in welded fabrication, it is possible to produce a practically fully rigid joint. Consequently, this type of connection is being increasingly used, particularly in high rise and slender structures, where stiffness requirements warrant the use of rigid connections.

10.4.1.3 Semi-Rigid Connection

Semi-rigid connection comes in between the two types mentioned earlier, and is designed to transmit shear force and a portion of bending moment across the joint. The basic principle of this type of connection is to provide partial restraint to the rotation at the beam end. However, the moment-rotation relationship of a semi-rigid connection detail is rather complex and there is no direct method to identify the end moment of such a connection; one has to determine the same by experiment conducted for the specific design or from the results of earlier experiments presented in specialist literature. This type of connection is not within the scope of the present text.

10.4.2 Design Considerations

While designing a connection for beams and columns, the designer must bear in mind the following considerations:

- The connection must comply with the requirements of the type of design used, namely, simple or rigid.
- The detail must be adequate to transmit the applied shear force and bending moment.
- The joint must be detailed to temporarily support the component members for welding during the erection process.
- The detail must provide for definite location of the joint to be established during erection.
- The joint must be as compact as possible.
- The joint must be fully accessible for welding.

10.4.3 Beam-to-Beam Simple Connection

As already stated, in a simple connection, the members are pin ended, that is, the connections are required to transmit shear only. Figure 10.14 shows two typical simple connections between a main beam and a secondary beam. In Figure 10.14a, the secondary beam rests on seating cleat which takes the shear reaction. The top flange of the secondary beam is cut short and its web is notched to clear the main beam. The web of the secondary beam is normally welded to the web of the main beam for keeping the beams at the joint in position. Although the weld is only nominal and does not extend to the full depth of the web of the secondary members, some bending moment is likely to develop. However, this moment is ignored in practice, since the main beam is weak in torsion and the joint would behave as pin-ended connection. In Figure 10.14b, an alternative (but similar) connection is shown. In this case also, the secondary beam rests on seating cleat which transmits the shear to the main beam, and the top flange of the secondary beam is cut short and its web is notched to clear the main beam. A web cleat is provided for keeping the beams at the joint in position and is nominally welded only at the toes to minimize fixity at the joint. The depth of the cleat should not

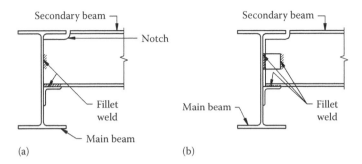

FIGURE 10.14
(a, b) Beam-to-beam simple connections.

be more than one-third of the depth of the beam. With this arrangement a reasonable degree of flexibility should be maintained at the joint. The seating cleat and the web cleat are shop welded to the main beam and the welds to the secondary beam are made at site.

10.4.4 Beam-to-Beam Rigid Connection

In this type of connection, the weld at the joint must develop the full moment acting on the joint. A few typical rigid connections between main and secondary beams are illustrated in Figure 10.15. In Figure 10.15a, the secondary beams are shallower than the main beam and therefore can be rested on two seating cleats shop welded to the main beam. The seating cleats must be of suitable size to accommodate the required length of weld to transmit the horizontal compressive force. The top flanges of the secondary beams and parts of the webs below are notched as shown and the flanges are butt welded directly to the edge of the main beam so that the top surfaces of both the main beam and the two secondary beams are on the same level. The vertical shears from the secondary beams are transmitted by the welds between the webs of the main and secondary beams. Figure 10.15b shows a similar connection but in this case the top flanges of the secondary beams and parts of the webs below are suitably notched so that the top flanges of the secondary beams rest on the top flange of the main beam and are fillet welded to it. Figure 10.15c also shows a similar connection as Figure 10.15a, but in this case the tensile forces at the top of the secondary beams are transmitted by a cover plate placed on top of the flanges. The length of the cover plate is dictated by the required weld length. The cover plate may be welded to the main beam. This weld however is only nominal. Figure 10.15d shows a detail where the main and secondary beams are of the same depth. In this case, the tensile and compressive forces are transmitted by cover plates and shear forces are transmitted by web cleats connected to the beams by weld (shop/site).

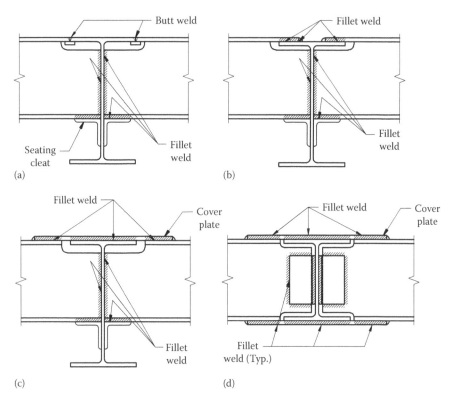

FIGURE 10.15
(a–d) Beam-to-beam rigid connections.

10.4.5 Beam-to-Column Simple Connection

In Figure 10.16, a few simple beam-to-column connections are shown. In Figure 10.16a, the beam is placed on a seating cleat, to which it is fillet welded. The top flange of the beam is connected to the column flange by a cleat. This top cleat is site welded on its *toes only* to allow for end rotation, which will ensure flexibility of the joint. The seating cleat is shop welded to the column and the weld is designed to take the end shear. For heavy reactions, the seating cleat will be replaced by a stiffened bracket, made up of joist cutting, rolled tee or plates. The weld between the bracket and the column shall be designed for vertical shear as well as bending moment due to eccentricity. In Figure 10.16b, a variation of the detail is shown where cleats on the top of the flange are replaced by cleats welded to the web near the top flange. In this case also the cleats are welded on the *toes only* to provide flexibility of the joint. In Figure 10.16c, a similar connection of a beam fitted into the web of a beam column is shown. In this case the seating bracket is made of a horizontal plate fitted into and shop fillet

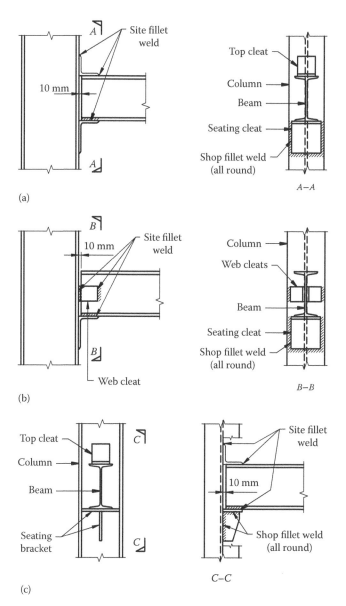

FIGURE 10.16
(a–c) Beam-to-column simple connections.

welded to the column web and flanges and stiffened by a vertical plate shop fillet welded to the underside of the horizontal plate and the web of the column. Needless to add the weld connecting the bracket and the column needs to be designed for vertical shear as well as bending moment due to eccentricity.

10.4.6 Beam-to-Column Rigid Connection

Rigid connections are required to develop the full moment acting on the joint with little or no relative rotation of members in the joint. As the beam flange forces (due to moment) are passed into the column, its web may require local stiffening to withstand the high shearing stress. Horizontal stiffeners are provided for this purpose. These stiffeners also serve to hold the column flanges and ensure uniform stresses in the connecting flanges.

Typical rigid connections between beams and columns are shown in Figure 10.17. In Figure 10.17a, the beam is located by a web cleat. The horizontal force in the bottom flange of the beam is transmitted to the column through a cleat, which also acts as a seating cleat and transmits the vertical reaction. The horizontal force in the top flange is transmitted through a connecting plate placed on top of the flange plate and fillet welded to the beam and full penetration bevel butt welded to the column. This welding will require a gap of 4 to 5 mm between the plate and the column. Normally, a backing plate is placed under the connecting plate to support the latter. The length of the connecting plate is dictated by the length of the fillet weld required to transmit the flange force. It may also be tapered in the thickness to facilitate fillet welding. Some authorities recommend to keep a length equal to its own width unwelded. If a long length of fillet weld is required, *U*-shaped slots may be cut at the end of the connecting plate for accommodating the weld. The stiffeners in the column web are fillet welded to the flanges and the web. For heavy vertical end shear, the seating cleat is replaced by a stiffened bracket. The weld connecting the bracket to the column flange shall be designed for vertical shear and bending moment due to eccentricity. Normally, the beam is kept short to allow for site adjustments. It may be noted that all site welds shown in this detail are downhand, which is an advantage in site operations. In Figure 10.17b, a rigid connection is illustrated in which the top connecting plate and the seating plate are dispensed with and the beams are directly welded to the column. The beam is connected with the column flange by profile fillet weld. The horizontal stiffeners for the column are similar to those shown in Figure 10.17a. In this detail, temporary seating cleat for holding the beams in position may have to be provided, which will be removed after the web of the beam has been welded to the column. This detail has several undesirable features. First, the length of the beam has little *margin* and does not allow for adjustments. Second, the detail includes overhead welding after erection, and makes greater demand on the welder and requires close supervision at site. Because of these drawbacks, this is not a popular detail, and should be avoided. Figure 10.17c shows a detail in which the beam frames into the web of a column. The beam is placed on a seat made of a horizontal plate with a vertical support plate underneath. The seating is, in effect, a stiffener to the column and fillet welded to it in the shop. The vertical plate is

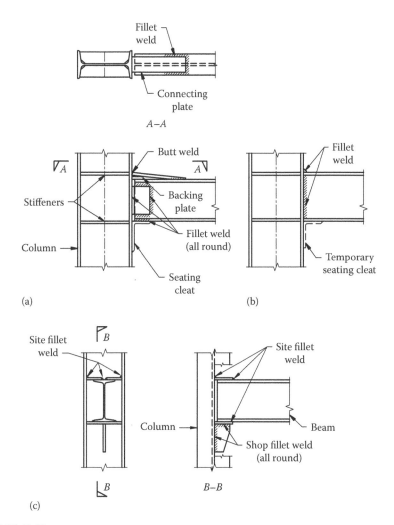

FIGURE 10.17
(a, b) Beam-to-column rigid connections.

also shop fillet welded. This seating system carries the vertical reaction, in addition to transmitting the horizontal force due to moment to the column. A pair of connecting plates are placed on top of the beam flange and are fillet welded to the beam flange and both the flanges and web of the column. The horizontal force at the top flange of the beam is transmitted to the column by this system of a pair of connecting plates. A pair of two plates is purposely used to provide longer length of fillet welds to transmit the required horizontal force.

10.5 Castellated Beam

Castellated beam is essentially an open web expanded beam and has been in use for many years. This type of beam is generally made economically by flame cutting the solid web of a rolled *I* beam in a zigzag pattern along its center line. The two parts are then taken apart and reassembled, so that the high parts of the zigzag line meet. The meeting edges are then welded. The result is a beam with a depth of 50% greater than the original, and also with considerable increase in the moment of inertia and the modulus of section. See Figure 10.18. This means not only higher carrying capacity but also smaller deflection compared to the solid web beam. The advantages are most pronounced for light loads on long spans where deflection is the criterion. The disadvantage of a castellated beam is the reduction of shear-carrying capacity of the web due to stress concentrations near the openings. Stress concentration can be reduced to some extent by carefully choosing the profile dimensions of the zigzag cut. Sometimes, it may also be necessary to *fill in* the end castellation to enhance the capacity of the web to carry the end shear. Similar detail should also be adopted in intermediate castellation where concentrated loads occur. In any case, where substantial concentrated loads are to be supported, castellated beams should better be avoided. Castellated beams are also not recommended for continuous beams across several supports.

Obviously, castellated beam is lighter than an equivalent solid beam with comparable properties. This produces saving in material and handling costs. Also, there is hardly any wastage in material. The web openings are often

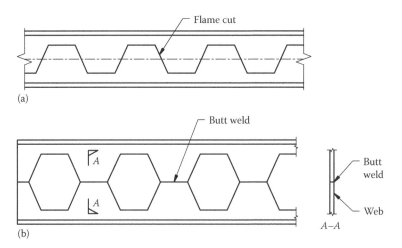

FIGURE 10.18
Castellated beam: (a) rolled beam web cut in zigzag pattern and (b) beam welded back to produce castellated beam.

used to accommodate service requirements such as piping and duct, which would otherwise be suspended below the beam. This arrangement indirectly provides savings in building height.

A tapered castellated beam can be fabricated by cutting the zigzag pattern along a preset angle to the beam axis and reversing one of the two halves and then welding together lengthwise in a similar manner as in the case of a normal castellated beam. Fabrication of castellated beam is a relatively easy process. Flame cutting of the web is normally done with a template-equipped machine. The joining process of the two halves of web is done by semiautomatic arc welding system. Full penetration square butt weld is provided with single pass on each side of the web.

Bibliography

1. Blodget, O.W., 2002, *Design of Welded Structures*, The James F. Lincoln Arc Welding Foundation, Cleveland, OH.
2. Dowling, P.J., Knowles, P.R., and Owens, G.W. (eds.), 1988, *Structural Steel Design*, The Steel Construction Institute, London.
3. Mukhanov, K.K., 1968, *Design of Metal Structures*, MIR Publishers, Moscow, Russia.
4. Owens, G.W., and Knowles, P.R. (eds.), 1994, *Steel Designer's Manual*, Blackwell Scientific Publications, England.

11

Plate Girders

ABSTRACT   The chapter begins with a brief introduction of the load resistance pattern of the components of a plate girder, namely, the two flanges and the web with special emphasis on welded connections. It describes typical weld details of shop and site splices for flanges and webs. Typical details of load bearing stiffeners, as well as intermediate stiffeners, are discussed with illustrations. The concept of postbuckling tension-field action and its use in design of plate girders are briefly discussed.

11.1 Introduction

Structural elements, subjected to moment and shear, usually take the form of either I- or box section for maximum structural efficiency. The cheapest form of I-section is a rolled steel beam. This is used for supporting normal loads over short spans. With the increase of loads and/or spans, the bending moments and shearing forces will also increase. In such cases, use of plate girder is the easiest alternative solution. In effect, a plate girder is a built-up beam consisting of two flange plates and a web longitudinally welded to form an I-section. A typical plate girder is shown in Figure 11.1. In a simply supported plate girder, the top and bottom flanges resist the axial compressive and tensile forces, respectively, arising from bending moment, while the web plate resists the shearing force. The transfer of the longitudinal shear from the web to the flange is done by the welds connecting the web to the flanges. For achieving maximum efficiency, the designer should aim at a section with least amount of welding compatible with the minimum weight of the component elements.

The primary advantage in the design of a plate girder over that of a rolled beam is that in a plate girder, the sizes of the primary components, namely, the top and bottom flanges and the web may be customized to the specific bending moment and shear dictated by the design for each situation, which is not always possible with the standard rolled beams. This customization of the cross section leads to an economy in the consumption of steel. However, a girder with too many variations in the cross section configuration may represent one with least metal, but may not be the most economical one, since

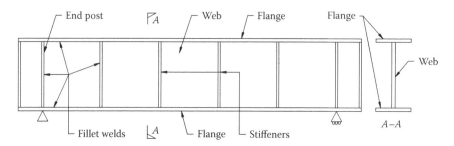

FIGURE 11.1
Typical plate girder.

excessive variation in plate sizes increases the fabrication cost. This aspect needs special consideration while finalizing the component sizes during design stage.

11.2 Flanges

In welded plate girders, each flange is generally designed to consist of a single cross section. However, in case the cross section needs to be varied due to design considerations, this can be done by adopting one or both of the following two methods:

- By changing the thickness
- By changing the width

11.2.1 Variation in the Thickness of the Flange

The thickness of the flange can be varied in two ways:

- By butt welding lengths of plates of different thicknesses in the transverse direction
- By fillet welding a cover plate on to the outer face of the flange plate

In the case of the first alternative, a series of plates of unequal thicknesses are laid end to end, and welded by full penetration butt welds to form a single length of flange plate. It is recommended that the thicker plate is chamfered down to the thickness of the thinner plate with a slope not exceeding 1 in 5 (see Figure 11.2). Where the thickness of the thicker plate does not exceed that of the thinner plate by 25% or is not more than 3 mm, the slope may be made up in the weld itself. Required radiographic test on the butt weld is to be done prior to the attachment of the flange plate to the web plate.

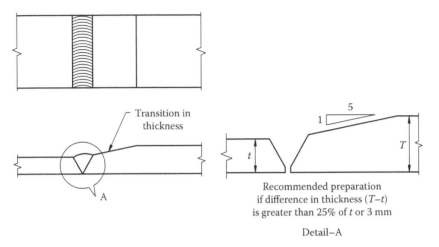

FIGURE 11.2
Method of transverse butt welding of plates of unequal thicknesses.

For short spans, use of transverse butt welded flange plates may not prove to be an economical solution, since butt welding is a costly process, and is likely to form a high proportion of the total welding cost. In such cases a second alternative, namely, use of welded cover plate may prove to be economical. Welded cover plate should be limited to only one on the outer face of any single flange plate. The fillet weld connecting the cover plate to the main flange plate should be designed to cater for the horizontal shear between these plates. In case any individual flange plate or cover plate is required to be joined to make up the length, the plate units are to be laid end to end and butt welded to form a single length. If the lengths of both the flange plate and the cover plate are made up by joining shorter units, butt welded joints should be staggered as far as possible.

The abrupt change in the flange cross section at the ends of the cover plate causes considerable stress concentration in the fillet weld in the region and makes it prone to fatigue-related cracking, typically initiating at the toe of the weld and propagating through the flange. Such cracks may even lead to total collapse of the girder. Examination of this condition is particularly important for design of crane girders, bridges, and similar structures, which are likely to be subjected to fatigue loading condition. In this context, two important conditions need special consideration. *First*, it will be expedient to ensure that the longitudinal force, which is evenly distributed across the cover plate, be transferred uniformly into the flange plate without causing any appreciable stress concentration. *Second*, reduction of the fatigue strength of the girder at the curtailment region should be avoided by reducing the cross-sectional area of the cover plate gradually, so that the reduction of the developed stress is also gradual. To avoid the possibility of such cracks,

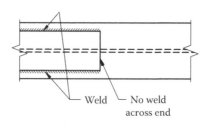

FIGURE 11.3
Detail at end of cover plate.

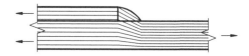

FIGURE 11.4
Large transverse fillet weld to avoid stress concentration.

many details for the ends of the cover plates have been suggested in different texts. Some of these are discussed here:

1. There should be no welds across the ends of the cover plate (Figure 11.3). However, in order to prevent ingress of water between the cover plate and the flange plate (to avoid rust formation) the ends need to be welded. Thus, the solution may not be attractive from the durability point of view. A suggested detail of end fillet weld for the cover plate is shown in Figure 8.10 of Chapter 8.

2. The first condition for consideration mentioned earlier, namely transferring the longitudinal force in the cover plate directly into the flange plate without causing appreciable stress concentration, may be achieved by providing a large transverse fillet weld (instead of a small fillet weld) across the ends of the cover plate as shown in Figure 11.4. This large fillet weld will help to transfer the force more uniformly through the fillet weld between the end of the cover plate and the flange plate, and thereby improve the fatigue strength of the joint by avoiding stress concentration.

3. The second condition for consideration, namely, *gradual* reduction of the cross-sectional area of the cover plate at the curtailment region can be achieved by providing a taper in the width of the cover plates at the ends. For a thick cover plate, the gradual transition of the sectional area may be improved by also tapering its thickness. However, adopting a tapered detail at the ends needs some further consideration. The presence of the web in this region makes the zone rather rigid, with little chance of localized yielding to prevent

building up of possible high stress concentration. A possible solution to guard against such a possibility is to adopt a large fillet weld (across the end portion of the tapered detail) as suggested in the previous paragraph (see Figure 11.4). This would help to transfer the force more uniformly and avoid high stress concentration. A typical detail showing suggested proportion for tapering is illustrated in Figure 11.5.

4. Some texts suggest that both the cover plate and the fillet weld at the ends should be ground to a taper to avoid fatigue related cracks. In fact, this detail has been incorporated along with tapering of the width of the cover plate in Figure 11.5.

5. Some authorities recommend the extension of the cover plate beyond the theoretical curtailing point to avoid the possibility of fatigue crack. The recommended extensions are as follows:

   a. Two times the nominal cover plate width for cover plates not welded across the ends.

   b. One and a half times the nominal cover plate width for plates that are welded across the ends.

   c. The cover plate should be extended beyond the theoretical curtailing point to within 1.5 m from the end of the girder.

While a number of alternative end details for cover plates have been discussed earlier, it is apparent that arriving at a decision in adopting a particular detail is not easy. The designer has to consider various aspects and balance the pros and cons of a specific detail he wishes to adopt.

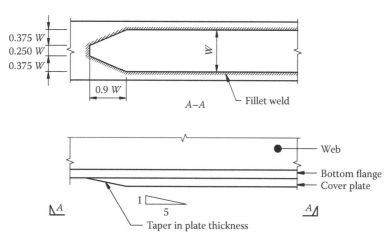

FIGURE 11.5
Suggested proportion for tapering of cover plate.

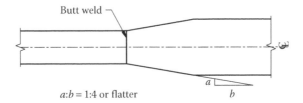

FIGURE 11.6
Suggested detail of transition in the width of flange plate.

11.2.2 Variation in the Width of the Flange

An illustration of smooth transition in the width of the flange plate is shown in Figure 11.6. However, the practice is not very attractive because of high fabrication cost.

11.3 Web

As in the case of flanges, the web area can also be varied by changing the thickness of the web according to the shears along the length of the girder. However, the saving in the cost of steel achieved by this method has to be balanced against the additional cost of assembly and welding, unless saving in weight of the steel is the critical factor. Thus, for short spans, due to economic reasons, a uniform thickness for the web is preferable. In longer spans, where joints in the webs cannot be avoided due to limits on the available lengths of plates or transportation consideration, varying the thickness of the web according to shear requirement may prove to be an economical solution. For cantilevers and continuous girders, variations in the web area is best achieved by varying the depth.

11.4 Web-to-Flange Welds

Sizes of fillet welds connecting the web to the flanges are determined by calculating the horizontal shear between the web and the flanges resulting from the bending forces on the girder. In some cases, these welds may also be required to carry vertical local loads on the flange and should therefore be designed accordingly. Where the web is machined and is in close contact with the flange before welding, such vertical loads from the top flange may be assumed to be transmitted in direct bearing from the flange to the web, provided that the bearing stresses are within the permissible limit. The load in such a case may be assumed to disperse at 30° to the plane of the flange. If the bearing stresses exceed the capacity of the plate, proper bearing stiffeners should be used to carry the concentrated load applied on the flange.

11.5 Transverse Stiffeners

In plate girders, two types of transverse stiffeners are generally used, namely, intermediate stiffeners and load bearing stiffeners.

11.5.1 Intermediate Stiffeners

Both horizontal and vertical shear stresses are set up in the web of a plate girder due to bending moments along the length of a plate girder caused by loads on the girder. These shear stresses combine and produce both diagonal tension and compression. In a deep and thin web, the diagonal compression could cause the web to buckle (see Figure 11.7). Intermediate transverse stiffeners are provided in such cases to increase the shear carrying capacity of the web plate and thereby prevent web buckling.

These also serve another important purpose. In a plate girder, the web is not an isolated plate—it is a part of a built up rigid structure (Figure 11.1). Thus, as soon as the critical buckling stress is reached, a new load carrying mechanism is developed, which prevents the girder to collapse. In this post-buckling scenario, while the flanges carry the bending moment, the buckled web serves as an inclined tensile membrane stress field (tension field) and the transverse stiffeners become vertical compression struts. Figure 11.8a illustrates the load carrying mechanism in the postbuckling range of the plate girder, which is similar to that of a Pratt truss shown in Figure 11.8b. The ultimate carrying capacity of the plate girder is considerably increased under this analysis. Most of the modern codes are based on this concept.

For the design and spacing of the transverse intermediate stiffeners, the stipulations of the governing code are to be followed. Intermediate stiffeners may be plates, angles, or tees and may be provided on either one or both side(s) of the web plate. Considering savings in both weight and fabrication cost, single stiffeners are economical, and there does not seem to be any valid theoretical objection to this arrangement. As in the case of load bearing

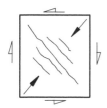

FIGURE 11.7
Diagonal compression from shear forces.

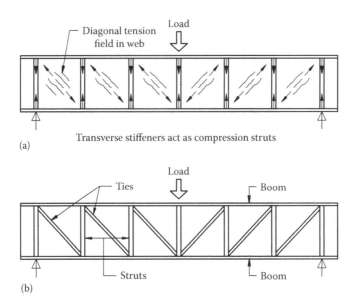

(a)

(b)

FIGURE 11.8
Tension field action: (a) tension field action in individual subpanels and (b) typical Pratt truss for comparison.

stiffeners the inside corners of the stiffeners should be suitably notched to allow the flange-to-web welds to be continuous.

11.5.2 Load Bearing Stiffeners

These are located at points of support and are designed to resist the end reaction as columns with an effective area, which includes a portion of the web. A few typical load bearing stiffeners are shown in Figure 11.9. The inside

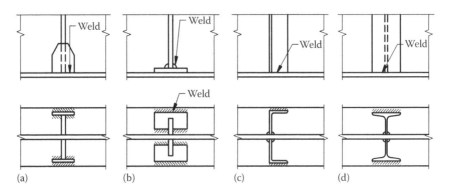

(a)   (b)   (c)   (d)

FIGURE 11.9
Typical details of load bearing stiffeners at tension flange.

corners of the stiffeners should be suitably notched to allow the flange-to-web welds to be continuous and uninterrupted and thereby avoid multiaxial weld concentration or intersection at a point.

11.6 Stiffener-to-Web Welds

The size of the fillet welds connecting stiffeners to web plates is governed by the thickness of the web or stiffener, whichever is greater. The topic is discussed in Section 8.3.2 of Chapter 8. Additionally, for load bearing stiffeners, the size of the weld should be adequate to transmit the full reaction or load to the web.

Some authorities allow intermittent welds for connecting stiffeners to the web. However, in structures such as crane girders and wind girders, which are subjected to frequent variations of stresses, intermittent welds should not be used to avoid the possibility of fatigue failure. Also, in locations vulnerable to moisture ingress and corrosion, this type of weld should be avoided. Where intermittent welds are used, the following requirements are recommended:

- Where fillet welds are placed on one side of the stiffener or on both sides, but staggered, or where single stiffeners are butt welded to the web, the effective length of each weld should not be less than 10 times the thickness of the stiffener.

- Where intermittent fillet welds are placed in pairs (one on each side of the stiffener), the effective length of each weld should not be less than four times the thickness of the stiffener.

- The distance between the effective lengths of any two welds in a line (even if staggered on opposite sides of the stiffener) should not exceed 16 times the thickness of the stiffener, nor 300 mm.

11.7 Stiffener-to-Flange Welds

11.7.1 Load Bearing Stiffeners

Load bearing stiffeners should be connected to the compression flanges by welds designed to take the full reaction or load, unless it can be assured that the connecting faces have tight and uniform bearing. However, it is generally recommended that welding across the flow of stress in the tension flange of dynamically loaded plate girders should be avoided since such welds lower

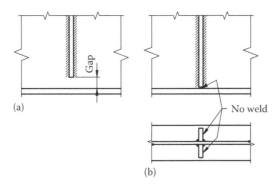

FIGURE 11.10
Typical details of intermediate stiffeners at tension flange.

the fatigue strength. Commonly used details with stiffeners for avoiding such welds are shown in Figure 11.9 and discussed here.

- A vertical plate is fillet welded on to the edge of the stiffener, which is connected to the flange by weld in the longitudinal direction. See Figure 11.9a.
- A small steel pad is inserted between the stiffener and the tension flange, with fillet weld between the pad and the tension flange in the longitudinal direction. Also, there is transverse fillet weld between the stiffener and the pad. See Figure 11.9b.
- With angle or T-section stiffener, the profile of the stiffener provides its own longitudinal plate for connection and the transverse welds are omitted (see Figure 11.9c and d).

11.7.2 Intermediate Stiffeners

Intermediate stiffeners are normally welded to the compression flange to provide proper support to the flange. It is a common practice for intermediate plate stiffeners to be stopped short of the tension flange, thereby avoiding transverse welding. Alternatively, the stiffener is tightly fitted on the tension flange to prevent ingress of moisture, and there should be no weld between the stiffener and the tension flange. These details are illustrated in Figure 11.10.

11.8 Splices

Two adjoining units of a girder are required to be spliced to enable them to function as an integrated unit. The type of splice depends primarily on whether it will be done in the shop or in the site.

11.8.1 Shop Splices

In case of shop splices, full penetration butt welded connection may be used. Where a clean and aesthetically pleasing appearance is required, butt welded site connections are the obvious choice. Full penetration butt welded splices produce sound connections, provided these are executed properly and inspected closely during execution. The resultant welded butt joints become at least as strong and as malleable as the parent material. Consequently, these joints are considered to be continuations of the plate itself, and in general, except for fatigue condition, do not call for separate design calculation. In detail drawings, full strength butt weld are specified without calculation. Selection of the type of butt weld for splicing generally depends on a number of factors. These include the following:

- Availability of suitable equipment for edge preparation
- Availability of suitable facilities or manipulators for turning over the work
- Size of electrodes available

For making up the required length of flanges or web of a plate girder, shop splices are almost always done with full penetration butt welded joints. These should be done preferably before the components are fitted together and welded. It is not necessary for these shop splices to be in a single vertical plane; rather, they should be located to suit the available length of the component plates or where a transition in section is required.

For thinner plates, single-V butt joints are generally used. For thicker plates, double-V butt preparations are most commonly used, since facilities are easily available in the shop for rotating or turning over the work for welding in downhand position or vertically. As has already been discussed (Chapter 7), a double-V preparation consumes less weld metal, and therefore, is more economical compared to a single-V butt weld. Also, since the welding is balanced, chance of angular distortion is minimal (see Chapter 4). Adequate back-gouging is necessary before the first run on the second side is commenced. In cases where downhand welding is limited to one side only, and where large size electrodes are available, a single-U preparation is an ideal solution.

11.8.2 Site Splices

Site splices for plate girders are mostly done by manual metal arc process. Consequently, it demands considerable knowledge and high degree of expertise on the part of the welders, the supervisory staff and the inspectors, in order to produce welded joint of acceptable quality. While finalizing the weld details, it is also imperative to ensure that the welder's task is made as easy and simple as possible. The adopted structural detail should be such as

to allow most of the welding work to be performed in the downhand position, which makes the job easier and faster. Adequate access for the stick
electrode to the required location should be ensured. Also, there should be
sufficient room for the welder's head and mask to enable him to see what
he is doing. In short, simplicity of the detail, ease of access for welding, as
also ease of inspection should be the primary objective in the designer's
approach. Essentially, in a site splice for plate girder there are three components to be spliced, namely, two flanges and one web. This can be detailed in
three ways as illustrated in Figure 11.11:

- Figure 11.11a shows the two flange splices and the web splice in the
  same vertical plane.
- Figure 11.11b shows the three splices in three different vertical planes.
- Figure 11.11c shows the two flange splices in the same vertical plane
  and slightly shifting the web splice.

Of these, the first alternative is most attractive for its simplicity. It has an
advantage in that it is easier to prepare the joints and maintain the fit up
properly when all the joints are in the same vertical plane (instead of being
staggered). Also, the web to flange fillet weld can be completed in the shop
itself, unlike the other two alternatives, where some site welding, including
overhead fillet welding, will be required to be done at site. Also, completed
shop fillet ends to the very end of the girder pieces, enables the two pieces to
be clamped together by draw cleats located at both the flanges and the web
for temporary supports during erection (see Figure 11.11a). In the case of the
other two alternatives, however, the flanges are not shop welded at the ends

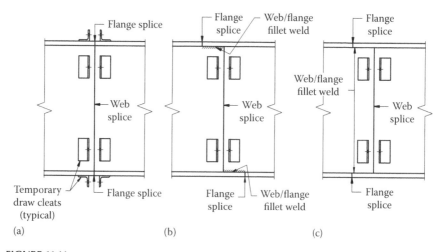

FIGURE 11.11
Alternative locations of welded flange and web splices in a plate girder.

and temporary draw cleats are attached in the web only (see Figure 11.11b and c). It is sometimes argued that splices, being weak points in the structure, should be staggered. In the case of butt welded splices, however, staggering of the splice locations does not, in any way, improve the performance of the girder. Therefore, locating all the three splices in the same vertical plane appears to be the most practical solution.

The web plates up to 12 mm thick are normally of single-*V* preparation. For webs over 12 mm thickness, double-*V* preparation may be preferred to balance any possible angular distortion. This also reduces the consumption of weld metal.

For splicing flange plates, which are normally much thicker than the web plate, either single-*V* or double-*V* preparation are used. Butt welds with single *V* have a tendency for angular distortion. Butt welds with double-*V* preparation would, no doubt, be an improvement by balancing the distortion, but would involve overhead welding. As already discussed, in site splicing, manual arc welding process is normally used, where downhand welding is preferred, being easier and faster than overhead welding. A compromise detail having an asymmetrical double-*V* preparation may be used. A typical proportion is having two-thirds of the flange thickness on the top and the remaining one-third on the bottom. This detail would reduce the amount of overhead welding and would also substantially balance the tendency of the angular distortion. Figure 11.12 illustrates these alternative preparations for butt welding.

In a butt welded splice, shrinkage of the girder components is inevitable, since weld metal contracts on cooling. For producing a good joint, good ductility of the metal is of vital importance. The other aspect that needs to be addressed in this connection is that shrinkage allowance in the length of flange and web plates must be provided while cutting these plates. As, for example, in a girder with different thicknesses of flange plates, because of greater number of weld runs, the amount of shrinkage will be more in the

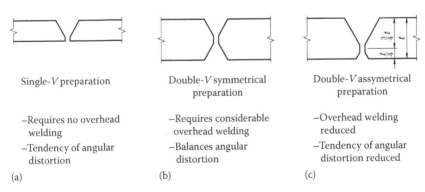

Single-*V* preparation	Double-*V* symmetrical preparation	Double-*V* assymetrical preparation
−Requires no overhead welding	−Requires considerable overhead welding	−Overhead welding reduced
−Tendency of angular distortion	−Balances angular distortion	−Tendency of angular distortion reduced
(a)	(b)	(c)

FIGURE 11.12
Alternative preparations for butt welded flange splices.

thicker flange plate. This aspect needs also to be considered beforehand. Otherwise, an unwanted camber may be developed in the profile of the girder after welding.

The sequence of welding should also be carefully selected in order to avoid problems. If the flanges are welded first, the initial gap between the webs of the two units should be larger than the gaps provided between the flanges, so that after the flanges are welded, even with their consequent shortening, the remaining gap between the webs is adequate for satisfactory welding. The thumb rule to be followed in this respect is to alternately weld the flanges and welds to the proportion of their depths. A suggested sequence is given here:

1. Weld full width of about half to one-third of the thickness of both flanges.
2. Weld full width about half the thickness of the web.
3. Complete the welding of the flanges.
4. Complete the welding of the web.

For deep girders, the web is usually divided into two or three sections and welding is done first in the top section, followed by the next lower section, and so on. The direction of the welding for each section should start from the bottom and proceed vertically up. In order to allow proper welding of the butt joints of the flanges, coped holes are generally provided in the web plates at those locations (see Figure 11.13). While this detail helps in butt welding the flange splices properly, it also tends to induce stress concentration at that location and thereby reduce fatigue strength of the girder. In order to counter this effect, the coped holes may be plugged by weld metal and ground flush after the butt weld has been completed and inspected.

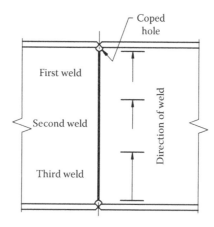

FIGURE 11.13
Sequence of welding for deep webs.

Bibliography

1. Dowling, P.J., Knowles, P.R., and Owens, G.W. (eds.), 1988, *Structural Steel Design*, The Steel Construction Institute, London.
2. Blodget, O.W., 2002, *Design of Welded Structures*, The James F. Lincoln Arc Welding Foundation, Cleveland, OH.
3. Xanthakos, P.P., 1994, *Theory and Design of Bridges*, John Wiley & Sons, New York.
4. Park, S.H., 1984, *Bridge Rehabilitation and Replacement*, S.H. Park, Trenton, NJ.
5. Brooksbank, F., 1952–1953, *Welded Plate Girders*, Quasi Arc, Bilston.
6. Tonias, D.E., 1995, *Bridge Engineering*, McGraw-Hill, New York.
7. Ghosh, U.K., 2006, *Design and Construction of Steel Bridges*, Taylor & Francis Group, London.

12

Portal Frames

ABSTRACT A brief introduction to the concept of a single-storey portal frame is given at the outset, followed by different types of portal frames and their construction methods. Common connections at knee and apex joints, rafter site joints and column bases are discussed.

12.1 Introduction

Portal frames are simple structures with clean lines and are one of the most common forms of construction of low rise single-storey structures. These frames are essentially rigid framed structures and comprise columns and rafters (horizontal or pitched) connected at the knees (and apex in the case of pitched structure) by moment-resisting connections. These connections are generally stiffened by suitable haunches (Figure 12.1). The basic form of this type of structure was developed during World War II.

Portal frames, being rigid, provide clear span that is unobstructed by bracing. They provide large column-free areas, thereby reducing the number of internal columns and their foundations. They can also support heavy concentrated loads.

The main disadvantage of portal frames is their susceptibility to differential settlement and temperature stresses. Also, foundations need to cater for horizontal reactions.

12.2 Types of Portal Frames

Portal frames can be either single or multiple bays and be either horizontal or pitched. These can have pinned or fixed bases. In the present text, single-span pitched portal frames will be discussed.

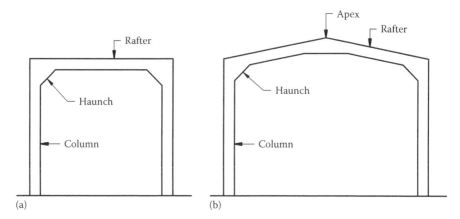

FIGURE 12.1
Typical single-span portal frames: (a) square corner portal and (b) pitched portal.

These frames may be constructed by

- Using standard rolled beam sections.
- Using rolled beam sections, but split diagonally down the web and then reversed and welded, resulting in a tapered beam configuration.
- Using beam sections fabricated by welding separate plates.

The most economical solution in the selection of the type of section to be used depends on various factors, namely, the span, shape, and the loading. Thus, for a simple portal of reasonable proportions, the use of standard rolled beam sections may prove to be most economical, because of minimum amount of welding required for fabrication, notwithstanding the fact that the strength of the section provided is required at only a few locations. Similarly, use of tapered beam configuration may save material, but would cost more due to cost of additional labor and welding in splitting beam sections and rewelding. However, in many cases, this type of construction has provided economic solutions. Fabrication of portal frames using sections built up from plates may prove to be economical for larger jobs, since the extra cost of welding for building up the sections from plates is likely to be offset by intelligent use of material to its maximum capacity at different locations. The details of joints illustrated in the succeeding sections may be suitably adapted for such cases also. Latticed portal frames using rolled angle sections or hollow sections are also popular, particularly in covering long spans, such as airport terminals and aircraft hangers.

Portal frames are typically used in spans in the range of 30–40 m, though spans up to 60 m or even 80 m have been achieved. Common bay spacing of portal frames is 6 m. However, this may vary between 4.5 and 10 m, the greater spacing being associated with the longer span portal frames.

The eaves height of common industrial buildings ranges between 4.5 m and 6.0 m. In the case of aircraft hangers and similar special structures, this height may go up to 15 m. Roof slope of portal frames is largely dependent on the wind characteristics vis-à-vis the configuration of the building and may be between 6° and 12°. Horizontal deflection, particularly at low slopes, should be critically considered and proper foundations to take care of the large horizontal thrust need to be provided.

12.3 Knee and Apex Joints

In a portal frame, one of the most important joints is the knee joint, where the stress effects are rather complex and need to be considered before the joint is detailed. The knee joint needs to be designed to cater for the following three conditions:

- Transfer of the end vertical shear from the beam to the column
- Transfer of horizontal shear from the column to the beam
- Transfer of end moment from the beam to the column

A typical member force resolution with diagonal stiffener is shown in Figure 12.2. The welds of the knee joint are to be designed to transmit the

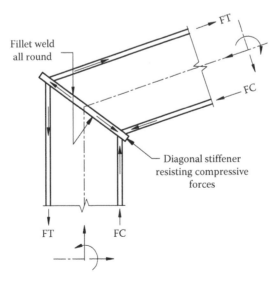

FIGURE 12.2
Typical force resolution with diagonal stiffener.

flange forces and web forces into the diagonal stiffener (division plate), which is designed as strut to take the compression forces. The division plate must be checked for laminations (during fabrication) as this will have little moment resistance if laminations are present.

12.3.1 Simple Joints

Simple knee and apex joints for a rolled steel beam portal frame are shown in Figure 12.3. In details 1 and 2, division plates of appropriate sizes are provided and are fillet welded all round to the connecting beams. This simplifies the fabrication. However, it must be ensured that no lamination exists in the division plates, since over some of their lengths, the plates will be in tension across their thicknesses. Also, the strength of the fillet welds requires special considerations in these details when the angles of the flanges with the division plates are too large or too small depending on the slope of the rafter. Alternative details 1a and 2a eliminate the problems encountered due to *too large* or *too small* angles between the flanges and the division plates. In these cases, bevel butt welds are used to connect the flanges to the division plates. The cost of preparation will, however, increase the overall cost of fabrication. Care must be taken to ensure that complete penetration is obtained in the bevel butt welds for which satisfactory sealing runs are obligatory, particularly at the inner flange connections, which are prone to stress concentration at the reentrant angles of the joints. In such locations, the division plates may be chamfered to facilitate deposition of the sealing runs as shown in the alternative detail 2a in the figure. In case the angles at the outer flanges are too small for providing satisfactory sealing runs, the welds to the outer flanges may be made with closed roots and the angles between the fusion faces should be made not less than 75° so as to allow full penetration of the butt welds. Such welds may not be strictly full strength butt welds, but for structures subjected to primarily static loading, these should be suitable. Transfer of the web loads is made via fillet welds and through the division plates. Another alternative connection type is shown in details 1b and 2b. In this type, the division plates are kept up to the inside faces of the outer and inner flanges of the beam sections, leaving gaps between the edges of the beams and the division plates to be filled in by weld metal to form butt welds. The fusion faces in the flanges may be splayed for ease of butt welding, in cases where the thickness of the division plate in relation to the depth of the weld is not sufficient.

12.3.2 Haunched Joints

For larger spans or heavier loading conditions, haunched joints are used. The greater depth of the haunched joint provided at locations of maximum bending moments increases the local strength and at the same time allows the basic section of the portal frame to be reduced. This leads to overall economy in the material cost. Figure 12.4 illustrates a typical example of haunched

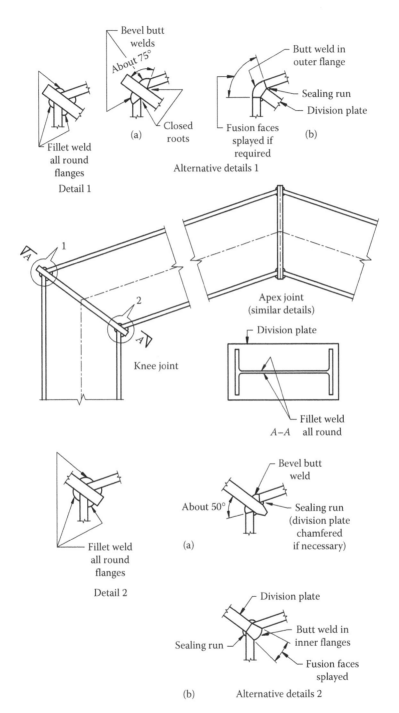

FIGURE 12.3
Simple knee and apex joints.

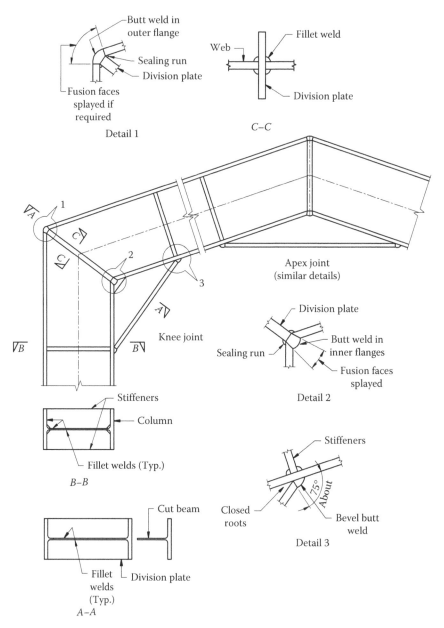

FIGURE 12.4
Haunched knee and apex joints.

knee and apex joints for a rolled steel beam portal frame. The connection details between the rafter and columns are similar to those shown for simple joints in the preceding section (see details 1 and 2). The haunches can be made from rolled tee sections or cutting rolled steel beam preferably with nontapered flanges. Alternatively, haunches may be made from web and flange plates welded to form tee section. Division plates are provided to act as diagonal stiffeners at the connections between the main frame members. It may be noted that by providing haunches, the loads on the inner flanges of the main beam at the location of the division plates are reduced. Stiffeners at the ends of the flanges of the haunches are provided to strengthen the flanges of the main beams against local bending and webs from buckling. Usually, the inner angles between the flanges of the main beams and the flange of the haunch are too small to permit satisfactory sealing runs (see detail 3). This problem may be resolved by making the welds with closed roots and the angles between the fusion faces about 75° to facilitate penetration at the root. In fabricating these joints, the haunch brackets should be added after the knee welding is completed. While there may be other forms of knee and apex joints, which can be detailed for portal frames, it is considered that those described are perhaps the most suitable, being both simple and economical to fabricate.

12.4 Rafter Site Joints

In the preceding sections, knee and apex joints have been designed for shop welding on the assumption that site joints would be provided at suitable locations in the rafters, where bending moments would be comparatively small. In practice, however, the location of site joint is generally finalized after considering the limitations for transporting the individual units to the site and erection facilities, rather than by the location of points of contraflexure. The frames are generally transported in three sections, namely, the rafter section and two frame legs. These site joints may be either bolted or welded. However, from an aesthetical point of view, these joints should be site welded, and every effort should be made to reduce the cost by carefully planning the welder's time on site. Generally, column units are erected first, followed by the rafter unit and the site joints are made at a height. For this purpose, the joints should be detailed so as to provide suitable landing surface for the connections to enable the rafter unit to land on the column units. Temporary cleats are normally provided to enable the joints to be aligned prior to welding. For very large frames, it may be necessary to subdivide the columns and rafter further. These sections are site welded to their full lengths and the column to rafter joints made in the usual manner after erection.

For small portal frames, it may be possible to lay the units on the ground and complete the welding prior to erection. In such cases, the rafter joints may be omitted and the knee joint made by partly shop and partly site welded. This method would not only be more cost effective, but would make possible a more accurate check on dimensions. Also, the single completed unit may be erected by one lift by a crane.

Typical welded site joints in the rafter are shown in Figure 12.5. In Figure 12.5a, four temporary angle cleats are provided to act as horizontal seating during erection. These cleats are to be removed when the flange welds and sufficient of web welds have been completed. Also, a landing flat may be tack-welded to the underside of the lower section, thereby dispensing with the need for temporary erection jigs. This flat is to be removed after the main downhand welding is completed and then the overhead sealing run is provided.

An alternative simple site joint is shown in Figure 12.5b. In this case, a division plate (wider than flange) is used and the beam ends are fillet welded all round to both faces of the division plate. With this system, the need for edge preparations of the flanges is eliminated. However, the division plate needs to be checked for lamination. The division plate is to be wider than the flange of the rafter beam and should be shop welded to the lower length of the rafter and site welded to the upper length. For erection purpose, two temporary angle cleats are provided. These should be removed when flange welds and sufficient web welds have been completed.

12.5 Bases

Portal bases comprise base plate (of the column), holding down bolts and foundation. The base plate is the steel plate arrangement at the base of the column, which is connected to the column by welding. Holding down bolts are the bolts that anchor the column base into the concrete foundation. Foundation is the concrete footing to resist compression, uplift, shear, and overturning moments as necessary. In the present text, the connections between column, base plate, and holding down bolts will be primarily discussed. Portal column bases may be pinned or rigid, depending on the detailing of the column base and anchor bolts. However, selection of the base type is normally governed by a number of issues. Thus, in view of the difficulty and expense for providing moment resisting rigid bases, theoretically pinned bases are very commonly used for portal frames, unless there are good reasons, such as the need to restrict deflections. However, care should be taken to detail the bases to be *erection-stiff*, so that the columns may be free-standing, with minimum guys or prop-supports during erection. For this purpose, the holding down bolts can be located entirely inside a line across the tips of the larger columns. For smaller columns, the

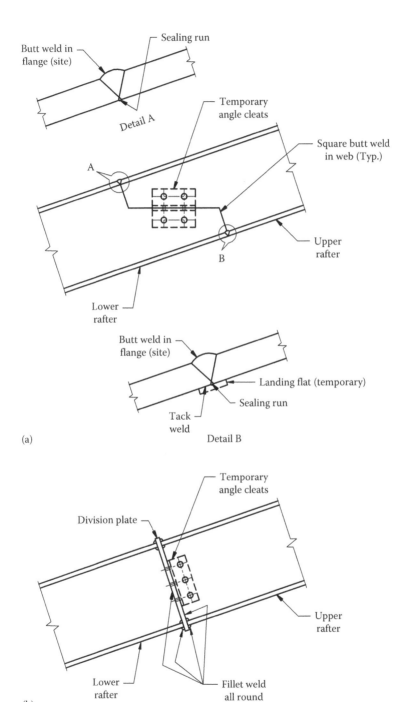

FIGURE 12.5
Typical welded site joints in rafter.

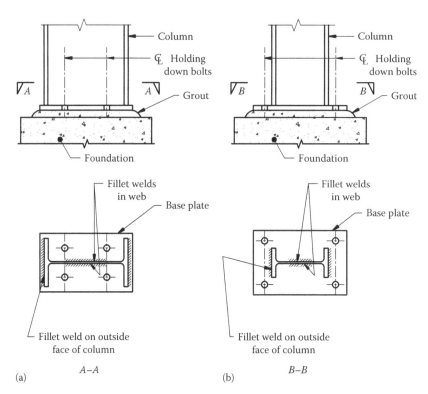

FIGURE 12.6
Typical nominally pinned-column bases: (a) for normal columns and (b) for small columns.

base plate may wider, so that the holding down bolts can be located outside the flanges. These are illustrated in Figure 12.6a and b, respectively. For a rigid moment-resisting base requiring bigger lever arm of the holding down bolts, a thicker base plate will be required. The column should also be fillet welded all round its profile to the base plate. If necessary, vertical stiffeners may be provided in the center line of the web of the column section to reduce the thickness of the base plate. The suggested arrangement is shown in Figure 12.7a. Additional gusset plate may be required when the moment is heavy as shown in Figure 12.7b. For more details of column bases the reader should refer to Chapter 10.

Generally vertical loading acting on the roof of the portal framed building produces horizontal shear forces acting outward at the base of the column. Horizontal reaction at the base may also be caused due to wind or any other horizontal force acting on the structure. The welds connecting the column to the base plate need to be designed not only for the vertical forces (downward or uplift) but also for the earlier-mentioned horizontal force.

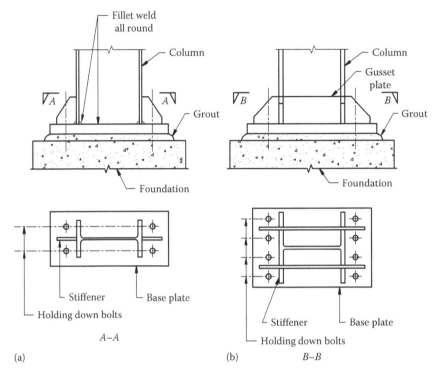

FIGURE 12.7
Typical rigid (moment resisting) column bases.

Bibliography

1. Blodget, O.W., 2002, *Design of Welded Structures*, The James F. Lincoln Arc Welding Foundation, Cleveland, OH.
2. Salter, P.R., Malik, A.S., and King, C.M., 2004, *Design of Single-span Steel Portal Frames*, The Steel Construction Institute, Ascot.
3. *Welded Details for Single-Storey Portal Frames*, 1957, British Constructional Steelwork Association, London.
4. Dowling, P.J., Knowles, P.R., and Owens, G.W. (eds.), 1988, *Structural Steel Design*, The Steel Construction Institute, London.
5. Owens, G.W., and Knowles, P.R. (eds.), 1994, *Steel Designer's Manual*, Blackwell Scientific Publications, Oxford.

13

Trusses and Lattice Girders Using Rolled Sections

ABSTRACT   The salient features of trusses and lattice girders, their typical usages, advantages, types, and characteristics, as also the analysis of primary and secondary stresses are generally discussed in the beginning. These are followed by various topics related to welded connections, namely, design methodology, design criteria, types of connections, design philosophy of internal joints, site splices, support, and bracing connections.

13.1 Introduction

Beams and plate girders supported on columns are useful for normal roof loads up to certain spans. Beyond such spans, where large column-free space is required for operational reasons, use of roof trusses along with lattice girders can be a convenient solution. Figure 13.1 illustrates an arrangement where roof trusses are supported by transverse roof girders (lattice girders) instead of columns, thereby providing a substantially increased column-free area.

A truss or lattice girder may be compared to a deep beam or plate girder where the flanges have been replaced by top and bottom chords and the solid web has been replaced by an open web composed of a triangulated framework of structural members. This type of girder is often termed as *open web girder*.

Theoretically, the members are connected to each other at multiple joints (panel points) by pins or hinges and the external loads are applied at joints. Consequently, members are subjected to only direct axial forces (tension or compression), leading to better utilization of the material than in a solid beam or plate girder. In actual practice, however, in a truss or a lattice girder with welded joints, the members are subjected to shear and moment also, producing secondary stresses, which are generally small and are ignored.

The terms *truss* and *lattice girder* are generally applied to a planer truss, where loads are applied only at the plane of the truss or lattice girder. For supporting loads applied in the other directions, a three-dimensional truss (space frame) is required. Such truss is outside the scope of the present text. Also, lattice girder used for bridges is not covered here. In effect discussions

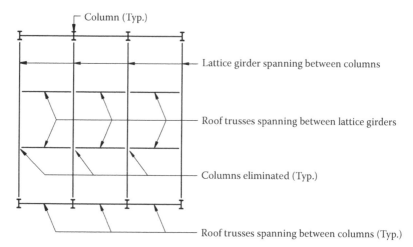

FIGURE 13.1
Arrangement showing roof trusses with lattice roof girders.

that follow cover single plane truss and lattice girder used primarily in roofs and in bracing systems forming lattice girders between structural members (e.g., columns) in a building.

13.2 Typical Usage

Trusses and lattice girders are used to support roofs and floors of a vast range of structures such as

- Workshops
- Buildings
- Sports complexes and stadiums
- Hangers for aircrafts
- Electric transmission line towers
- Radio transmission towers
- Framework for overhead traveling cranes
- Lighting towers
- Observation towers

Additionally, truss system is also used to provide stability in buildings where bracing members form a truss system with columns and other structural members (see Figure 13.2).

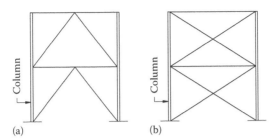

FIGURE 13.2
Column bracings: (a) K bracing and (b) cross bracing.

13.3 Advantages of Welded Roof Truss

In spite of the increasing popularity of welded portal frame, welded roof truss is still the most widely used type of roof structure. This may be attributed to its lightness, as well as to the simplicity in its design. Also, it hardly needs any special or complicated equipment for its fabrication. A small fabrication shop can produce a satisfactory truss for use in factory buildings and storage sheds.

Design of a truss as welded structure in preference to riveted/bolted one presents a number of other advantages also. Some of these are as follows:

- No deduction in the section for holes is required in tension members, resulting in a reduced size of these members and thereby saving the weight of the truss. However, this saving may be partially offset in the case of members subjected to reversal of stress.

- For smaller trusses, gusset plates may be eliminated for majority of the connections, thereby achieving further saving. However, it may not, as a rule, mean saving in weight, since sufficiently larger sections need to be selected for the main members, so that enough space is available for welding the connecting secondary members.

- Considerable fabrication cost in marking and drilling of holes is eliminated.

- Smaller sections can be used for secondary members, as size of the section is no longer dictated by rivet back marks and edge distances. Within limits, the sections may also be made thinner.

- One other point in favor of welded joints is that they are aesthetically more pleasing than bolted joints.

In this connection, two points need to be mentioned. First, end preparations of the secondary members may cause higher fabrication cost. Second, special jigs and fixtures required for setting up and assembly of welded trusses are rather costly, and therefore, unless repetition of similar trusses is ensured, the economic advantages may be reduced considerably. In spite of these disadvantages, however, the overall cost should show advantage with welded truss and lattice girder.

13.4 Truss Types and Characteristics

Trusses and lattice girders are categorized in respect of their overall shape (e.g., triangular, trapezoidal, parallel chord, and single pitch), as well as the internal arrangement of web members (e.g., Pratt, Howe, and Fink). Some of the most widely used types are shown in Figure 13.3.

In the case of Pratt truss under normal vertical loading, the (longer) diagonals are in tension while the (shorter) verticals are in compression (with

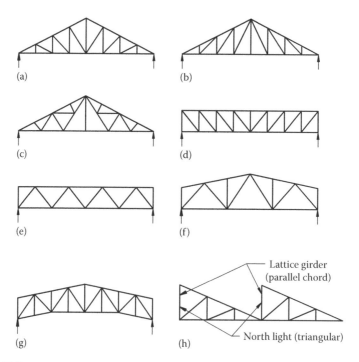

FIGURE 13.3
Common types of roof trusses and lattice girders: (a) Pratt (triangular), (b) Howe (triangular), (c) Fink (triangular), (d) Pratt (parallel chord), (e) Warren (parallel chord), (f) modified Warren (trapezoidal), (g) combination of Howe and Pratt (partial parallel chord), and (h) north light (triangular).

lower slenderness ratio), resulting in economy in the sections adopted. However, in such an arrangement, the forces in the compression chords at the apex are comparatively higher than the force in the tension chord at the mid-span, thereby reducing the advantage gained in the design of web members. In a comparatively low-pitched Pratt roof truss, the reversal of force pattern will put the longer diagonals into compression, thereby needing heavier sections.

The force pattern in the web members and the chords for a Howe truss is converse to that in a Pratt truss. Thus, a Howe truss can be advantageous where the normal vertical loads are low and where reversal of load due to wind will occur.

For a long span high-pitch roof, Fink truss may be the most economical solution, as the compression chords are subdivided by web members into shorter lengths, ensuring lower slenderness ratio of these chords. The designer may use his discretion in arranging and subdividing the chords and web members to make the truss most economical. A note of caution to the designer needs to be made in this context. Introducing too many members will increase the fabrication and maintenance costs, which are likely to offset a portion of the advantage gained in the steel weight.

For parallel chord Warren trusses, lengths of compression members are generally identical to those of the tension members. This configuration results in considerable saving in the fabrication cost, as also, reduces the steel weight.

13.5 Analysis

13.5.1 Primary Stresses

The statics of simple truss or lattice girder is fundamentally based on the assumption that the neutral axes of all members intersect at node points, *which act as pins*. In the early days when trusses were made of wrought irons, pins were actually used for connecting the different members at the intersection points. With the advent of rivets and bolts, single-pin connections are replaced by more rigid joints by way of groups of rivets or bolts connecting the members to gusset plates at the intersecting points. The rivets or bolts are located at the back marks of the angle sections instead of at the neutral axis lines of the connecting members, thereby introducing secondary stresses. It was, however, found that the effect of these (more rigid) joints is sufficiently small and may be ignored. Thus, in a riveted/bolted truss the forces in the members can be safely determined by simple statics. The welded joints in a light truss are comparatively more rigid than those with bolts. However, in the case of a welded truss (unlike in a riveted/

bolted truss), the neutral axis lines can be used for setting out. This system compensates for the extra rigidity of the connection to some extent. Thus, in the analysis of welded truss when the neutral axes of different members meet at node points, the joints may be assumed to act as pins, in which case the structure is subject to primary stresses only.

13.5.2 Secondary Stresses

Secondary stresses in the planer truss are caused by moments produced by certain conditions, which are discussed here.

13.5.2.1 Loads Applied between Intersection Points

Location of purlins away from intersection points of top chord is the most common example of this condition. In such a case, bending stresses on the top chord, considered as a continuous beam, are to be taken into account along with the axial forces while designing the top chord.

13.5.2.2 Eccentricity at Connections

During design stage, it must be ensured that the center lines of the elements converging at a nodal point intersect at its center. With this arrangement, the forces converging at the nodes will balance and the members will be subjected to tension or compression only. If the center lines do not intersect at a point, extra moment at the joint will be created leading to secondary stress, which must be avoided. Figure 13.4a shows the correct method of layout where the center lines meet at a point, while Figure 13.4b shows a detail with eccentricity, which should be avoided. The aim is to ensure that the center lines should coincide with the center of gravity of the section, or at least be as close to it as possible.

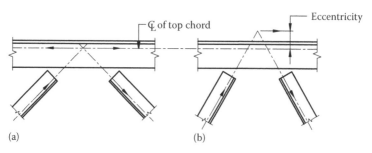

FIGURE 13.4
Layout of members at nodes: (a) concentric and (b) eccentric.

13.5.2.3 Joint Rigidity and Truss Deflection

In general, for trusses used in buildings, where imposed loads are predominantly static, effects of such moments are negligible and are generally neglected.

13.5.2.4 Torsional Moment

Secondary stresses may also be produced due to torsional moments introduced by members not acting in the plane of the truss, such as monorail girders hanging from the bottom chord of a truss. In such cases, members may also need special attention during design for fatigue effects, in which case failure may occur at lower stress levels than those at which failure would occur under static load.

13.5.3 Rationale of Analysis

For rigorous elastic analysis of rigid jointed, redundant, or continuous trusses and frames with loads between intersections, several computer programs are available, which are commonly used now for analysis of frames with considerable accuracy. However, care must be taken to adopt consistent approach for analysis and design of the frame. Thus, in case the secondary stresses are to be ignored in the design, the primary forces must be analyzed, assuming the truss as a pin jointed frame. In case a rigid frame analysis by a computer is used, the effects of moments due to secondary stresses are to be considered along with the axial forces obtained using a computer.

13.6 Connections

13.6.1 Design Methodology

Design methodology of connections of a truss or lattice girder may be briefly described as follows:

1. Calculate the forces in a particular joint and identify the correct load paths.
2. Based on that, make a preliminary sketch showing the proposed layout.
3. Compute the length of weld required from forces already calculated and arrange them suitably in the layout sketch. This should be done

considering maximum efficiency of the connection, making it as compact as possible to economize on use of material.

4. Finalize the layout sketch, considering practicability for fabrication, and ease of execution.

13.6.2 Design Criteria

In the development of welded connection design, the criteria that need careful consideration are briefly discussed here:

1. The connection should be as direct and as simple as possible. It should be complete and structurally in equilibrium, and should satisfy the requirements of the governing codes.

2. The design of the connection should be in line with the analysis of the truss or lattice girder.

3. Behavior of local connections (e.g., monorail connection in the bottom tie), which participate in the functioning of the connection, needs to be considered and appropriate details introduced to make the connection strong and durable.

4. As already discussed, care should be taken to ensure that the centroidal axes of the converging members in the nodes meet at a single point, so as to avoid secondary stresses in the joint. In cases where the centroidal axes do not meet at a point, the eccentricity should be considered in the design of the connection and also of the concerned members. This aspect should be taken into account while finalizing the form of the truss configuration.

5. Details in connections should avoid sudden change in sections as they are likely to become the focal point of stress concentration. This aspect is important in structures subjected to alternating loads leading to fatigue fracture.

6. The overall size and degree of squareness of a shop fabricated component are always subject to manufacturing tolerances, which include mill tolerances for the sections and plates, as also fabrication tolerances allowed in the governing codes. These aspects should be carefully considered while detailing a connection.

7. Design of connection should consider ease of fabrication, as well as easy access for inspection during fabrication, erection, and maintenance of the structure during its service life.

8. Possibility of accidental damage during handling, transportation, and erection at site should attract the attention of the detailer during finalization of any connection.

13.6.3 Types of Connections

In an open web girder system, connections may be broadly classified into four categories.

13.6.3.1 Internal Joints

These joints usually connect the web members (verticals and diagonal members) to continuous chords.

13.6.3.2 Site Splices

Fabricated units sent from the shops are assembled and spliced at site to make up the completed structure. Very often these are located at the node points.

13.6.3.3 Support Connections

These are required for the open web girders to transmit the load to the support structures at the ends of the girders.

13.6.3.4 Bracing Connections

These are generally required to connect the lateral bracing systems of the vertical open web girders, columns, and other structural members.

13.6.4 Internal Joints

13.6.4.1 Transmission of Forces in Chords

There are two distinct types of force transmission in chords.

1. Where the chord is continuous through the gusset, the force in the chord is directly transmitted within the chord itself, and only the difference of chord forces is transmitted through the gusset. Welded connection is to be designed accordingly.

2. If chord splice is required, this is sometimes made outside the joint. This arrangement is normally used in heavier trusses in order to relieve the gusset plate of large forces. Where the chord splice is located right at the node point, the gusset is subjected to comparatively heavy forces, since the gusset along with the splice cover members, transmit the entire force from one chord to the other. The centroid of the splice cover material should be made to coincide, as far as possible, with that of the chord member. In such a case, it may be assumed that the cover materials act as one and in line with the chord. The centroid of the gussets, on the other hand, is usually

some distance away from that of the chords and consequently the gusset plates are subjected to bending. For light trusses, this effect may be ignored; however, for heavier trusses, gusset plates should be checked for such bending. For heavier trusses, when the free edges of the gusset plates are subjected to compression, these should be stiffened by welding suitable plates near the free edges.

13.6.4.2 Connection Arrangements between Main Members and Web Members

For welded trusses and lattice girders, connections between the main members (rafters, chords, ties, etc.) and the web members are normally done either by welding the web members directly to the chords or using gusset plates. These are discussed here.

13.6.4.2.1 Connection without Using Gusset Plates

In the case of light trusses of welded design, gusset plates are generally eliminated, and single members are used for main members to enable secondary members to be welded to the main members directly. Use of tee sections for both main and secondary members has a theoretical advantage, since detailing can be so arranged as to make the truss symmetrical about a vertical plane running through the center of the tees. Figure 13.5 illustrates a typical arrangement at the joint. In this case, the flanges of the secondary members are slotted so as to allow them to fit over the stem of the main tee section. The stems of the secondary members are also cut to fit snug on the stem of the main member. Both butt weld and fillet welds are used to connect the members. However, preparations necessary for the ends of the tee sections are rather expensive and consequently restrict the use of this arrangement.

An alternative arrangement with tee sections for main members and double angles for web members is quite satisfactory. A typical arrangement is shown in Figure 13.6. If necessary, a gusset plate may be butt welded directly on the edge of the stem of the main member to accommodate longer length

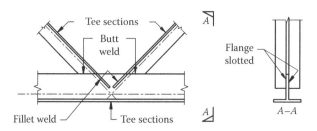

FIGURE 13.5
Typical joint with tee sections.

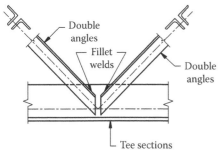

FIGURE 13.6
Typical joint with double angle web members.

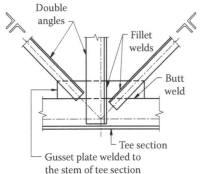

FIGURE 13.7
Typical joint with welded gusset plate.

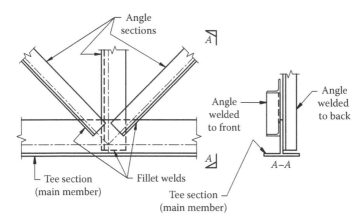

FIGURE 13.8
Typical joint showing secondary members placed both in front and back of main member.

of fillet welds. The weld may be ground flush in locations where the secondary angle members are connected (Figure 13.7).

In order to simplify the detailing of the joints, often some of the secondary members are placed at the back of the main members and others in front. A typical detail of such an arrangement is shown in Figure 13.8.

The effect of the transverse eccentricity, thus created is not seriously detrimental to the structure and therefore may be neglected in the design calculations.

13.6.4.2.2 Connections Using Gusset Plates

For heavier roof trusses, it is convenient to have the main members to be made up of two angle sections in the form of a T-shape with gussets welded in between. The secondary members may be single or double angles depending on the forces being transmitted, and are connected to the gusset plates by welds. The following aspects related to connections using gusset plates need special consideration:

1. *Transport and erection:* In order to avoid deformation of trusses during transportation and erection, the aspect of stiffness of the truss unit in the direction perpendicular to the plane of the truss needs to be carefully addressed. Use of unequal leg angles for selected members with their longer legs outstanding from the plane of the truss may be considered for this purpose. However, availability of the variety of standard equal leg angle being considerably greater than that of unequal leg angles largely offsets the advantage gained by the use of unequal leg angles.

2. *Ease of fabrication:* For easy fabrication, it is a common practice to cut off the web members at right angles to the center line. Also, care should be taken to leave 30 to 50 mm space between the ends/edges of structural members so as to provide access for welding the same on to the gusset plates.

3. *Gussets:* The forces in different members in a joint are transmitted to each other by means of gusset plate. Care must therefore be taken to ensure that the shape and the thickness of the gusset are designed to transmit the concerned forces without overstressing any section of the gusset. The shape and thickness of the gusset plate are dictated by

   a. Orientation or layout of the members meeting at the joint

   b. Weld length required for each member

   c. The magnitude of forces in the members, which need to be transmitted to each other via the gusset plate

Gusset plates connecting the chord members to the web members should be secured to the former by fillet welds on two sides of the chord, as shown in Figure 13.9a. The force in the chord will be transmitted to the gusset by all the four lines of welds. However, it is not always possible to adopt this detail. For example, the gusset cannot be protruded beyond the top face of the top chord at locations of purlin cleats. In such cases, the gusset plates are typically kept about 5 mm shy from the top surface of the chord

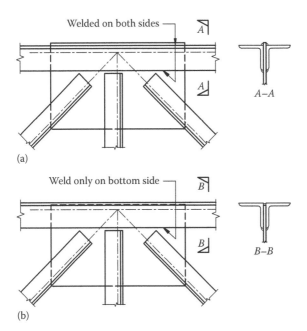

FIGURE 13.9
Connection of gusset to chord: fillet welded on (a) both sides and (b) only on bottom side of chord.

angles and are connected by fillet welds provided only at the edges of these angles as shown in Figure 13.9b. The space between the backs of the angles and top of the gusset plate is generally filled up with weld. This weld, however, is not to be considered for calculation of the strength of the joint. Only the welds applied at the edges of the chord angles are to be considered to transmit the force in the chord center line. Consequently, the welds will be subjected to additional bending stresses. Generally, these bending stresses are not large and some authorities recommend that such welds may be designed for shear only with a design capacity suitably reduced by about 15%–20%.

Generally, the thickness of gusset plates throughout the truss is kept uniform. Also, this is governed by the thickness of the members connected, as well as the minimum thickness as per codal provision. Typically, the thickness of the gusset plate is to be checked against the following conditions:

- The gusset must be capable of resisting the maximum force from each web member across the critical section through the gusset (*a-a* in Figure 13.10).
- The gusset must resist shearing by the combined coexistent forces in all the web members across the longitudinal section of the gusset

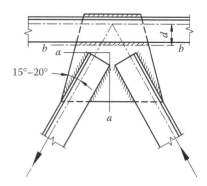

FIGURE 13.10
Design of gusset plate.

immediately beyond the weld line on the chord members (*b-b* in Figure 13.10). For simplicity, this shear *Q* can be taken as the sum of the horizontal components of the maximum forces in the web members.

- The gusset must also withstand the bending moments due to the maximum shear force *Q*, the bending moment at any section being *Q* multiplied by *d*, where *d* is the distance from the center of inter-section of the joint to the section considered. The critical section will probably be *b-b* in Figure 13.10.

- In certain cases where although the centroidal axes of members con-verge at a point, the resultant force does not lie on the center line of the gusset plate, resulting in bending moment, for which the gusset should be checked.

13.6.4.3 Spacer Plates

In trusses and lattice girders with gusset plate connections, double angle main and web members should be joined together in the panel space between the gussets by means of spacer plates as shown in Figure 13.11. This is particularly required for compression members, where one of the angles may buckle independently of the other one due to half the force in the double angled member, since the minimum radius of gyration of a single member is considerably smaller than that of a double angle.

Recommended maximum distance between the spacer plates is approxi-mately 40 *r*, where *r* is the radius of gyration of one angle in the plane parallel to the plane of the spacer plate. It is common practice to provide spacer plates in tension members also to secure uniform stressing of the two angles. The spacing of spacer plates for tension members may be lim-ited to 80 *r*.

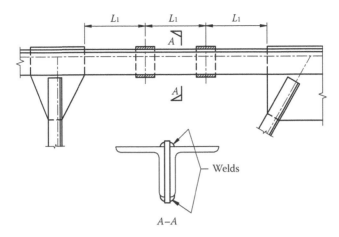

FIGURE 13.11
Typical arrangement of spacer plates.

The thickness of the spacer plate should be identical to that of the gusset plate. The plates are usually made 60–100 mm wide.

13.6.5 Site Splices

It is preferable to fabricate the entire truss or lattice girder in the workshop and transport the same as a single unit to site for erection. However, often, it becomes necessary to break up the structure into smaller units in the workshop due to one or more of the following situations:

- Limitations in the available length of the rolled section
- Limitations in the transportation facilities
- Ease of handling of erectable units with the available erection facilities

In such cases, the part units of the structure are required to be assembled and joined at site with the help of suitably designed splices. Splicing at site may be bolted or welded. Some authorities prefer bolted splices because of ease of assembly leading to speedier operation and economy. Also, in structures subjected to alternating stress, bolted connections are widely used because of improved performance in fatigue conditions. However, welded site splices are often used from an aesthetic point of view. Also, in case girders where gusset plates are altogether eliminated, site welded connections for assembly at site become obligatory. In all cases, the sectional area of the splice material should be provided in accordance with the requirements of the code being followed.

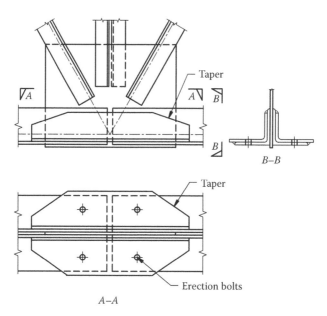

FIGURE 13.12
Typical bottom chord splice with tapered splice angles.

Splices may be located either at the nodal points or in the panel. The former has an advantage since a part of the gusset area can be used to contribute to the required area of the splice. In welded splice angles, it is a good practice to taper the legs of the splice angles in order to avoid abrupt change in a cross-sectional area and consequently concentration of stress. Example of a typical bottom chord splice with gusset plate contributing to splice area with tapering splice angles is shown in Figure 13.12.

13.6.6 Support Connections

Roof trusses may be supported on steel columns, roof girders (commonly lattice girders), concrete columns, and brick walls. A typical example of a roof truss supported at the end on column, where forces meet at the nodal point, is illustrated in Figure 13.13. The dimensions of the bearing plate are dependent on the compressive strength of the support material. The bearing plate is welded on to the vertical end gusset and is connected to the column by (anchor) bolts and nuts. A typical end connection of a lattice girder to a column is shown in Figure 13.14. In this case, the lattice girder is rigidly connected to the column, and consequently will have a fixed end condition resulting in a support moment. It will be noted that two sets of stiffeners have been welded in the column—one at the top and another at the bottom chord levels of the lattice girder to strengthen the column web in resisting

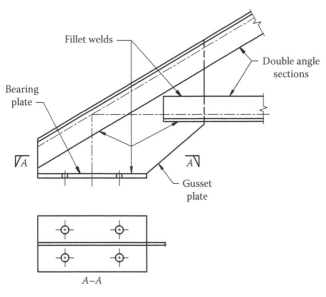

FIGURE 13.13
Support connection of truss.

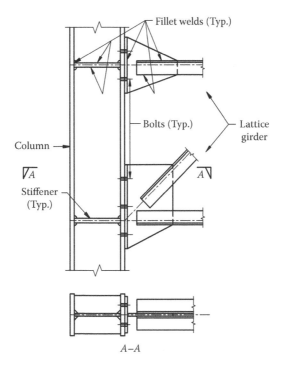

FIGURE 13.14
Rigid connection of lattice girder to column.

local buckling. The welded connections for these are to be designed for the horizontal forces due to the support moment. The welds connecting the gusset plates of the lattice girder and the bearing plates transfer to the column the horizontal force due to support moment, as well as the end shear force of the lattice girder. These are to be designed accordingly.

13.6.7 Bracing Connections

Bracing connections are generally required where bracing members form a truss system to connect vertical open web girders, columns, and other structural members to provide stability both during construction stage and during service period. Bracings are additionally useful during construction stage to facilitate plumbing and squaring of the building. Bracings are also designed to resist external horizontal forces such as wind, as well as internal ones such as braking forces of cranes. Bracings are normally provided in the plane of top and bottom chords of trusses and also in the vertical plane between trusses and columns.

There are two basic configurations for bracing system, namely, single diagonal type and cross member type as illustrated in Figure 13.15. In the first system, the members and connecting welds are to be designed for compressive as well as tensile loads. In the case of cross member system, however, only the members in tension are assumed to be effective and the members and welds are designed accordingly, while members in compression are checked against slenderness criteria only. Since reversal of loads is possible, all members and connecting welds are to be designed for tension and checked for slenderness ratio. Cross member system of bracings is the most popular system adopted in industrial structures. A typical connection of bottom chord bracings in a truss of lattice girder is shown in Figure 13.16.

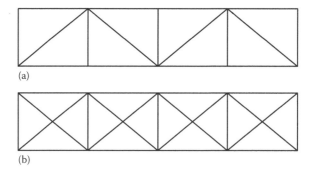

(a)

(b)

FIGURE 13.15
Bracing configurations: (a) typical single diagonal bracing system and (b) typical cross member bracing system.

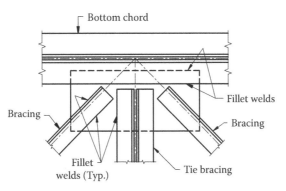

FIGURE 13.16
Bottom chord bracing connection.

Bibliography

1. Blodgett, O.W., 2002, *Design of Welded Structures*, The James F. Lincoln Arc Welding Foundation, Cleveland, OH.
2. Mukhanov, K.K., 1968, *Design of Metal Structures*, MIR Publishers, Moscow, Russia.
3. Ghosh, U.K., 2006, *Design and Construction of Steel Bridges*, Taylor & Francis Group, London.
4. Xanthakos, P.P., 1994, *Theory and Design of Bridges*, John Wiley & Sons, New York.

14

Trusses and Lattice Girders Using Hollow Sections

ABSTRACT The chapter begins with the historical background of the use of hollow sections in structures, followed by examples of their current usage and advantages. It then describes typical connections, their characteristics, structural analysis, design parameters, local stress distribution at joints, and various failure modes. Other aspects discussed are factors that control joint capacity, recommendations for butt and fillet weld details, and economy in fabrication.

14.1 Introduction

Use of hollow sections in structures is not a new concept to man. These sections have been in use as construction material in the past for a long time in the form of bamboo, a natural product. Even today, bamboo is used extensively in roof trusses, purlins, and posts for permanent and temporary structures in many developing countries in Asia. Bamboo is also used as temporary scaffoldings around buildings in many countries.

Use of wrought iron as an economically viable building material became apparent to structural engineers in the first quarter of nineteenth century. In a surprisingly short time thereafter, in 1850, Robert Stephenson built Britannia Railway Bridge over the Menai Strait, North Wales, using wrought iron as the construction material, applying an entirely new concept, namely, box girder—another form of tube or hollow section. This was followed by two other bridges, built by Isambard Kingdom Brunel, one across the river Wye at Chepstow, United Kingdom (1852) and the other, the Royal Albert Bridge over the river Tamar at Saltash, Plymouth, United Kingdom (1859), to carry railways, where tubular wrought iron sections were used as the main compression members. The next major structure using tubular sections was the Forth railway bridge near Edinburgh, Scotland, which was designed by John Fowler and Benjamin Baker, using steel as the construction material. Completed in 1890, this bridge had compression members comprising of 3.66 m diameter tubular sections. The earlier examples illustrate that some of the greatest engineers of yesteryears did appreciate the properties of tube or hollow sections as a potential structural form, which are now being gainfully utilized by their successors.

Coming to more recent past, circular steel tubes have been in use for quite some time albeit in a comparatively smaller scale in the bicycle and motor cycle industries. For many years, tubes were being joined by brazing process. Welded joints were introduced during the middle of twentieth century. However, fatigue cracks plagued the new technology and designers became skeptical in its use. Extensive researches carried out by the aircraft industry also helped to broaden the knowledge about the behavior of welded joints in hollow sections made up of aluminum and steel.

Circular steel tubes became very popular as structural material since the middle of the twentieth century. Of late, rectangular hollow sections have also gained considerable ground for their economy, strength-to-weight ratio, and aesthetic advantages. Researches sponsored by the steel industry worldwide for their promotion have led to an increased understanding of the behavior of welded joints in hollow sections. This understanding has led to the publication of several standards in many countries (e.g., the United States, European countries, Australia, Japan, and India) to regularize weld designing and welding procedures to ensure quality control in production to meet the requirements of the construction industry. As a result, these sections are now being used in many projects, such as bridges, buildings, and offshore structures in many countries.

While the individual members of the structural frame are selected and checked carefully to ensure safety and economy, often, enough attention and thought are not given to the design of the connections of the individual members, as to how they connect together, and whether they are adequate to transmit the forces to each other. It must be understood in the design stage itself that these welded joints are an integral part of the structure and are vital components in holding the individual members together. The other aspect that needs due consideration at the concept stage is that the capacity of a joint in a structure using hollow sections is predominantly determined by three parameters, namely, the member size, steel grade, and the joints geometry (slope). Since these are decided at the concept stage, it becomes imperative that the designer of the frame structure should also specify and design the joint. However, it is very often noticed that this important task is left to the contractor to be done during fabrication stage, when it may be too late to change member size, steel grade, or the geometry of the joint. It is therefore important that the designer is aware of the effects of the above parameters on the capacity of the joints from the earliest stage of concept design. It is in this context that this chapter deals with the various aspects of welded joints in hollow sections, so that these can be considered by the designer at the initial concept stage itself. The present text deals with joints mainly in plane frames using hollow sections under predominantly static loading conditions.

14.2 Typical Examples

Currently hollow sections are being used in a vast range of structures. Some of the more common uses are indicated here:

- Sports complexes and stadiums
- Airport terminals
- Hangers for aircrafts
- Electric transmission line towers
- Observation towers
- Radio transmission towers
- Roof trusses
- Railway electrification supports
- Bridges
- Framework for overhead traveling cranes
- Crane jibs
- Lighting towers

14.3 Advantages

Steel hollow sections are generally more expensive than open section profiles (such as I, H, channel, and angle sections) on per ton basis. However, a properly selected structural configuration using hollow sections, if designed with proper care and considerations, should almost always be light in terms of overall weight than a similar construction using open sections.

In the design of compression members, circular hollow sections generally have an advantage because of their uniform values of radius of gyration in all directions, unlike open sections, which have dissimilar values of radius of gyration in different axes. Also, being enclosed sections, the torsional properties of hollow sections are considerably better compared to those of open sections.

In addition, in case of circular hollow sections, the codes allow for reduction of exposed area (shape factor) in case of wind load, leading to further economy in construction and erection costs. Indirectly, this would also lead to significant reduction of loads and moments in the foundation design, thereby affecting further economy in the overall costs.

Also, hollow sections, being enclosed sections are less likely to be affected by weather hazards, if properly sealed at ends to prevent corrosion inside.

This is a major advantage from the durability point of view. There is a further saving in the material for corrosion (or fire) protection because of lower exposed area for these sections.

Added to these would be simplicity in welded connections, resulting in reduced fabrication, and erection costs. The cumulative effect of all these would generally result in a comparatively cost-effective structure.

Apart from economic viability, structures with hollow sections present a very pleasing appearance and enhance the overall aesthetic appeal of the structures. Consequently, these sections have, nowadays, become very attractive alternative construction material to the architects.

14.4 Types of Hollow Sections

Structural hollow sections are available in three broad categories:

- Circular hollow sections
- Rectangular hollow sections
- Square hollow sections

Selection of a particular category of hollow section to be used in a structure is to be made at the initial design stage. It is more often than not that aesthetics get preference over ease of fabrication or economics or even ease of joint capacity calculation, while selecting the category of the hollow sections.

14.5 Material Quality

Structural hollow sections are produced almost always from weldable quality steel. It is necessary for the designer to specify the appropriate code in this respect.

14.6 Connections

For welded lattice girders and trusses, connections between the main members (chords, rafters, ties, etc.) and the web members may be done either by traditional gusset plates or by directly connecting the members. In the first option, gusset plates may be welded on to the main members. The other

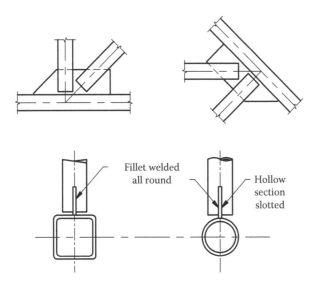

FIGURE 14.1
Typical connections with gusset plates.

connecting members are slitted at ends to suit the vertical gusset and welded all round to the gusset. The thickness of the gusset plates and weld length may be designed in accordance with the requirements of the governing code. Typical details of gusseted joint are shown in Figure 14.1.

Alternatively, the web members may be directly welded to the chord for transmitting forces. In such cases, member sizing has a direct implication in the detailing of the joint, as well as on the cost of fabrication. It is thus important for the designer to have prior knowledge about the significance of the design decisions on the joint capacity, as well as fabrication, assembly, and erection of the structure.

Broadly, connection may be classified into five categories: *T*, *Y*, *X*, *N*, and *K*. The category to be adopted in a particular connection is generally dictated by the geometry of the frame and its basic dimensions. Figure 14.2 shows these typical connection categories.

It is obvious that for square or rectangular hollow sections, the flat faces enable simple straight cuts for the connecting members, whereas for circular hollow sections, these cuts become complex as they need to match the profile of the connecting members. This would require proper templates produced by a computer for manual cutting (flame or plasma) or special profile cutting machinery.

In case of partially overlapping joints, it becomes difficult to provide all round weld for every connecting member end. Provisions of the controlling code should be followed in such cases. In case of fully overlapped joints, the toe of the overlapped web member should be welded satisfactorily to the chord.

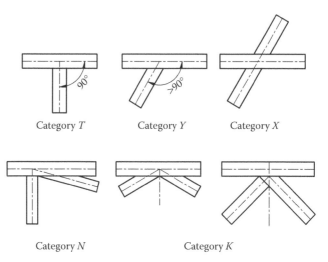

Category *T*   Category *Y*   Category *X*

Category *N*   Category *K*

FIGURE 14.2
Typical connection categories.

14.7 Structural Analysis and Design Parameters

Traditionally, lattice structures are analyzed on the basis of pin-jointed frame with the member center lines meeting at node points and loads applied at these node points. Thus, theoretically, the members are either in tension or in compression. However, the structures using hollow sections are almost always fabricated by connecting most of the members by welding and making site connection either by welding or by bolting. Also, the main members (ties, rafters, and chords) are normally continuous with web members connected to these main members by welding directly or via gusset plates. The welded connections introduce bending moments into the members due to their inherent stiffness. Thus, the connection system adopted for hollow sections virtually negates the concept of pin-jointed connection. Also, due to reasons of detailing, it may be necessary to introduce gap joints or overlapped joints for connecting the members, and thereby deviating from the intended idea of common meeting points for individual members and introduce eccentricities in the joints (see Figures 14.3 and 14.4).

However, notwithstanding the earlier mentioned anomalies, the concept of common (nodal) meeting points with pin-connections is normally accepted as this makes the analysis of the structure quite straightforward. The results of axial forces in the members obtained by this concept are also somewhat of good approximation for sizing of the members. The forces in

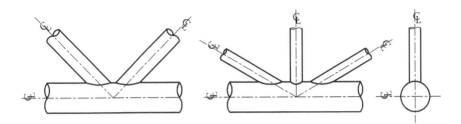

FIGURE 14.3
Member center lines meeting at node points.

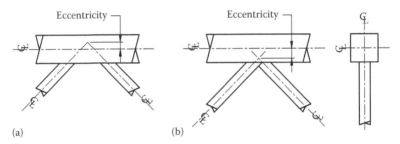

FIGURE 14.4
Joints with eccentricities: (a) gap joint and (b) overlap joint.

the members and joints resulting from these deviations are termed *secondary forces* and are required to be considered in the final design of the joints and members.

14.8 Local Stress Distribution

A typical local stress distribution at the joint of a welded square hollow section structure is shown in Figure 14.5. It will be noted that the stress pattern is not uniform. Therefore, welds need to be designed to cater for such nonuniformity of stresses. Also, the capacity of the fillet welded joint should not be less than the design capacity of the related member.

Satisfactory welding of the toe of inclined web member of the chord should be ensured in all cases. In case the inclination is less than 60°, the connecting web member end should be beveled and a butt weld provided (Figure 14.6), even if the remainder of the peripheral weld is fillet weld. In case the inclination is 60° or more, the weld type used for the remainder of the weld may be used for the toe also, that is, either fillet or butt weld.

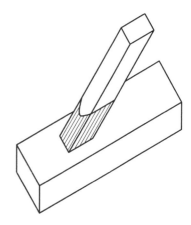

FIGURE 14.5
Typical local stress distribution at joint.

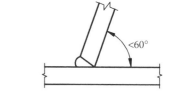

FIGURE 14.6
Typical butt weld for inclined connecting member.

14.9 Joint Failure Modes

Six recognized failure modes in joints of hollow sections are illustrated in Figure 14.7. These are as follows:

14.9.1 Chord Face Deformation

This is a case of the plastic failure of the chord face and is the most common mode of failure for joints with a single web member, and for K- or N-joints with a gap between the web members, when the web member to chord width ratio is less than 0.85 (see Figure 14.7a).

14.9.2 Chord Side-Wall Buckling/Yielding

This failure usually occurs when the web member to chord width ratio is greater than about 0.85, particularly in joints with a single bracing (see Figure 14.7b).

14.9.3 Chord Shear

This failure may occur in K- or X-joints in rectangular chords when the width of the section is more than the depth (see Figure 14.7c).

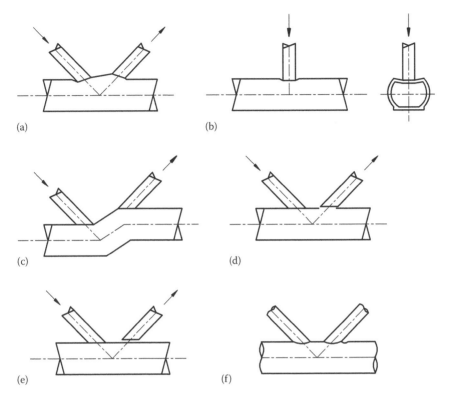

FIGURE 14.7
Failure modes in joints: (a) chord face deformation, (b) chord side wall buckling/yielding, (c) chord shear, (d) chord punching shear, (e) web member failure, and (f) localized buckling.

14.9.4 Chord Punching Shear

In this mode of failure, crack is first initiated in a square or rectangular chord leading to punching through the chord. Such failure occurs when the chord width to thickness ratio is small. For limiting ratio, relevant code may be referred (see Figure 14.7d).

14.9.5 Web Member Failure

This failure is primarily caused by nonuniform stress distribution across the web member, resulting in insufficient cross-sectional area of the web member and local overloading, leading to crack in the weld or the web member. This failure generally occurs in chord joints with large web member to chord width ratios and thin chords (see Figure 14.7e).

14.9.6 Localized Buckling

This failure may occur both in chords or the web members due to nonuniform stress distribution at the joint (see Figure 14.7f).

14.10 Joint Capacity

Capacity of the welded joint in a lattice-type construction broadly depends on the following two factors:

- The type of joint (single web member, double web members with a gap, or an overlap)
- The type of loading (axial forces in the components or moment in the joint)

In order to improve the capacity of a joint, the following aspects need special attention:

- It is generally economic to fabricate girders with gap joints than overlap joints.
- Lattice girders with larger chord members (compared to the web members) are likely to have improved lateral stability. Also, the larger size of the compression chord increases its radius of gyration, and consequently its capacity.
- Joint capacity (except fully overlapped joints) increases if smaller (but thicker) chords are used instead of larger and thinner ones. Similarly, larger thinner web members give stronger joints than smaller thicker ones.
- Overlapping joints are likely to possess higher strength than joints with gaps. Capacity of joints with gap is, however, increased if the web member to chord width ratio is increased. This criterion leads to selection of large thin web member and small thick chords.
- In case of partially overlapping joints, higher capacity is achieved if both the chord and the web member are kept small and thick.
- In case of fully overlapping joints, higher capacity is achieved if the overlapped member is as small and thick as possible irrespective of the size and thickness of the chord.
- A joint is stronger if the thinner member at a joint is welded to the thicker member, rather than *vice versa*.

These interacting variables need careful consideration at the time of selection of member sizes.

14.11 Joint Reinforcement

When a joint fails to satisfy design requirement, the easiest option is to change the deficient member size. But this option may increase the overall weight of the structure. The other option would be to modify the joint configuration. In case this is also not possible, it may be necessary to reinforce the joint locally. This can be done by welding a plate of appropriate thickness onto the deficient portion of the section.

For gap joints, a few illustrations of reinforcements by adding plates have been shown in Figure 14.8 with recommended dimensions. While selecting reinforcing plate, care should be taken to ensure that the plate is free from any lamination defect. The thickness of the reinforcing plate should not be less than the thickness of the parent member. Also, properties of this plate should be equivalent to those of the parent section or of higher grade. The welding should be all round the reinforcing plate.

For overlap joints with chords made up of square or rectangular hollow sections, a suggested detail showing reinforcement plates is shown in Figure 14.9. The reinforcement plate should be wider than the web member, to provide space for fillet welds with throat thickness not less than the thickness of the web member.

14.12 Typical Joint Details

Basically, both butt and fillet welds are used for connecting hollow sections. Sometimes, combinations of these are used such as in the connection of circular web member to a circular chord member, where the diameter of the former is 2/3 of the diameter of the latter or more. These aspects are discussed in the following paragraphs. Recommendations of the present text should be supplemented by the requirements of the governing code of practice for the specified project.

Butt welds are used when two hollow sections are required to be joined end-to-end, either in the same alignment or intersecting at an angle. The edges of the members may either be square or be a *single V* prepared depending on the thickness of the section, the angle of the intersection, the welding position, and the size and type of electrode used. Projection of the finished weld over the surface of the joining members (weld reinforcement) should not exceed 10% of the throat thickness of the weld. Alternatively, the projection may be ground flush. It is recommended to have the edges of the members prepared for thicknesses above 5 mm; that is, square butt welds should be limited to 5 mm thickness only. However, square butt welds beyond 3 mm thickness should have backing members. It is, however, recommended

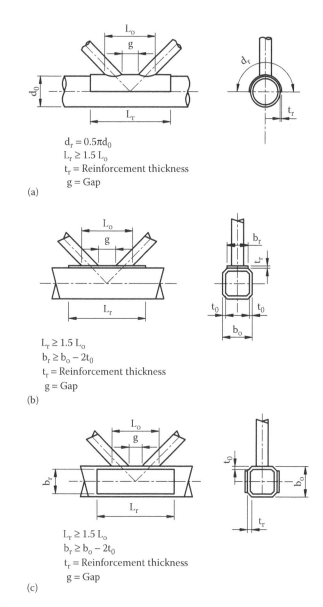

$d_r = 0.5\pi d_0$
$L_r \geq 1.5\, L_0$
t_r = Reinforcement thickness
g = Gap
(a)

$L_r \geq 1.5\, L_0$
$b_r \geq b_0 - 2t_0$
t_r = Reinforcement thickness
g = Gap
(b)

$L_r \geq 1.5\, L_0$
$b_r \geq b_0 - 2t_0$
t_r = Reinforcement thickness
g = Gap
(c)

FIGURE 14.8
Typical joint reinforcements in gap joints: (a) chord saddle reinforcement, (b) chord face reinforcement and (c) chord side wall reinforcement.

to provide backing members to all end-to-end butt welds. This would help in keeping the alignment of the connecting members and also ensuring sound root run and full penetration of the weld.

Figure 14.10 shows typical details of butt weld preparation, normally used for joining two members of the same size and thickness up to 20 mm.

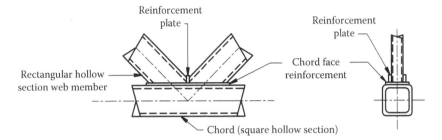

FIGURE 14.9
Typical joint reinforcements in overlap joints of square hollow section and rectangular hollow section members.

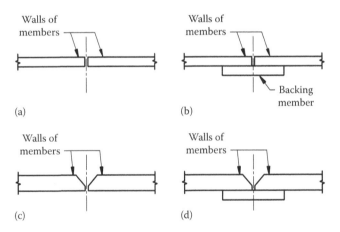

FIGURE 14.10
Typical details of butt weld preparation: (a) square butt weld (without backing member), (b) square butt weld (with backing member), (c) single-V butt weld (without backing member), and (d) single-V butt weld (with backing member).

For dimensions of the preparations governing code should be followed. For thicknesses above 20 mm, suitable welding procedure should be developed by trial and testing.

Butt welds in hollow sections are considered to have developed the full strength of the parent section provided the choice of electrode employed is correct and full penetration has been achieved. When the thickness of two members placed end-to-end vary, the external faces normally remain flush, and the change in the wall thickness occurs in the inside faces of the section. When such members are butt welded, the profile of the weld in the interior location should be smooth and not abrupt. This is necessary to avoid stress concentration in the weld, particularly in structures with dynamic loading

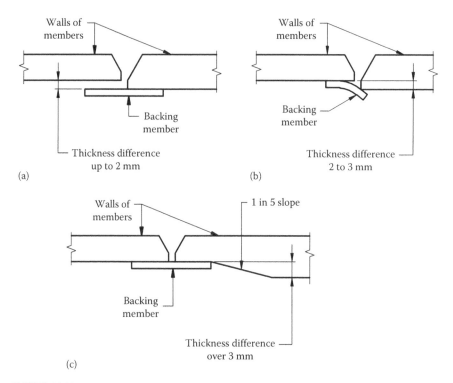

FIGURE 14.11
Typical backing member details for joints with different wall thicknesses.

(bridges, crane jibs, etc.), which may trigger fatigue cracking. In such cases, the following steps are recommended (see Figure 14.11).

- If the difference in thickness does not exceed 2 mm, the backing member should be tack welded to the thicker member. See Figure 14.11a.
- If the difference is above 2 mm, but below 3 mm, the backing member should be tack welded to the thinner section and bent by locally heating to rest on the inner face of the thicker member. See Figure 14.11b.
- For difference in thickness exceeding 3 mm, the inside face of the thicker member is to be machined, so that the backing member seats snugly on the modified inner face of the thicker member, as shown in Figure 14.11c. Backing members should be of same material as the parent section or equivalent. The size should be 20–25 mm wide and 3–6 mm thick strips.

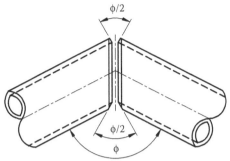

$\phi/2$

$\phi/2$

ϕ

FIGURE 14.12
Weld preparation for components meeting end-to-end at an angle.

When the two members meet end-to-end at an angle (as in the case of segmental arch or bowstring girder), edges of the walls are often cut at an angle equal to half of the intersecting angle (Figure 14.12). Backing member for such butt welds needs to be specially fabricated.

Butt welds are generally used for joining web members to main chords of structures using hollow sections. Two types of details are used for connecting web members to main chords. These are as follows:

- When the main chord is of circular section, the end of the web member is to be cut to a shape to match the curvature of the circular chord member and the angle of intersection between the web member end and the surface of the circular main member changes continuously around the perimeter of the web member end. The connecting end of the web member should be prepared to give, as far as possible a 45° single bevel between surfaces of the circular main member and web member. Figure 14.13 illustrates the detail.

- In the second type of joint, when the main chord is a rectangle or square hollow section, there is no continuous change in the angle of intersection between the web member end and the horizontal surface of the main member, and edge preparation becomes rather simple. Figure 14.14 illustrates the detail.

For ease of welding, the angle of intersection θ of the axes of hollow sections in both the joint types should preferably be kept at 30° or above.

Alternative details showing fillet welded connections of web members to the main chords are shown in Figures 14.15 and 14.16. Generally, fillet welding provides an economic answer to most of the joints in structures subjected to static loading.

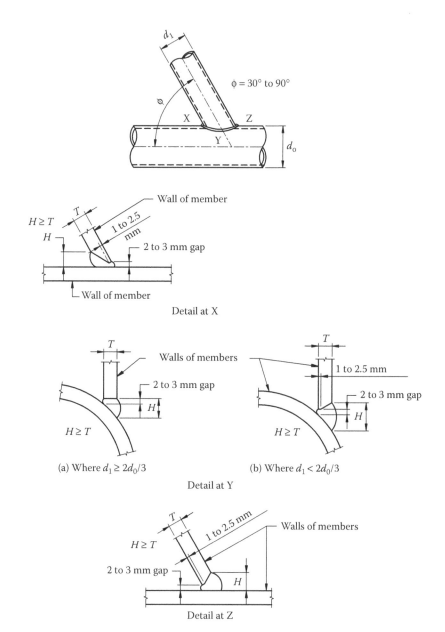

FIGURE 14.13
Typical details of butt welded joints for circular hollow sections. *Note*: For smaller angles, adequate throat thickness must be ensured.

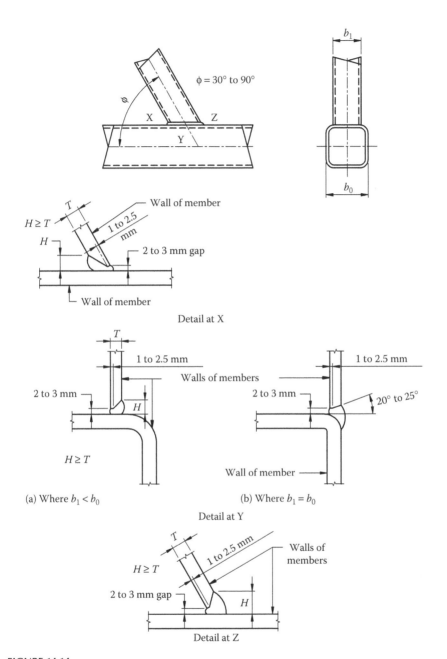

FIGURE 14.14
Typical details of butt welded joints for square and rectangular hollow sections. *Note*: For smaller angles, adequate throat thickness must be ensured.

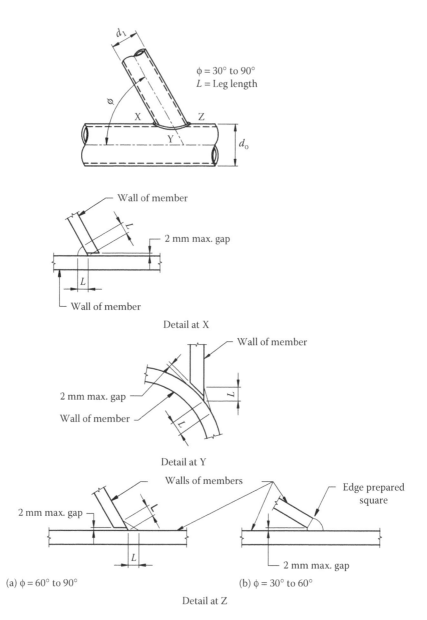

FIGURE 14.15
Typical details of fillet welded joints for circular hollow sections. *Note*: For smaller angles, adequate throat thickness must be ensured.

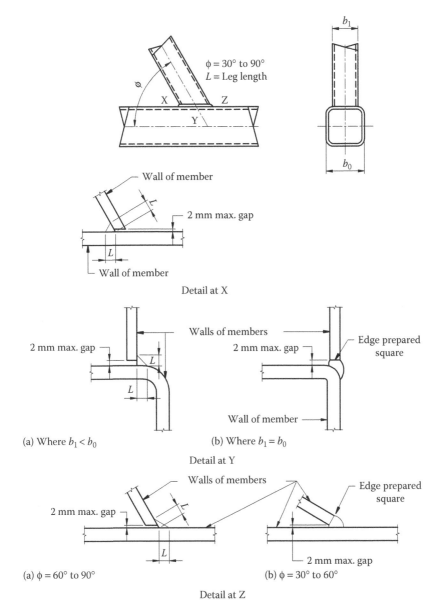

$\phi = 30°$ to $90°$
L = Leg length

Wall of member

2 mm max. gap

Wall of member

Detail at X

2 mm max. gap

Walls of members

2 mm max. gap

Edge prepared square

Wall of member

(a) Where $b_1 < b_0$

(b) Where $b_1 = b_0$

Detail at Y

Walls of members

Edge prepared square

2 mm max. gap

2 mm max. gap

(a) $\phi = 60°$ to $90°$

(b) $\phi = 30°$ to $60°$

Detail at Z

FIGURE 14.16
Typical details of fillet welded joints for square and rectangular hollow sections. *Note*: For smaller angles, adequate throat thickness must be ensured.

14.13 Economy in Fabrication

In a triangulated configuration made up of hollow sections, the ease of fabrication and consequently the economy in fabrication depends largely on a few factors.

First, the ease with which the chords, as well as the web members, can be put into position and then welded at the joints will determine the cost of fabrication. Obviously, joints with the chords with gap between the web members have comparatively more tolerance for fitting up and would be easier to assemble and weld, compared to overlapping joints. In case of overlapping joints, particularly in partially overlapping joints, tolerance to fit up is often minimal for achieving the desired panel point location. Also, in the latter case, welding will be more time consuming.

Second, end preparation and welding of individual members in lattice girder form a large percentage of the overall cost. Therefore, it is always expedient if the number of web members is kept to the bare minimum. Girder with web members of Warren or N-type configuration is likely to be more cost effective.

Third, choosing the type of hollow section—circular or rectangular or square—will also affect the cost of fabrication. In case of circular hollow section girders, ends of each member will be required to be cut to a shape to match the curvature of the connected member. In case of overlapping joints, the cutting to proper shape becomes all the more labor intensive. If the chords are made of square or rectangular hollow sections with web members of any type of hollow sections, the labor involved is drastically reduced as only a single cut is required at each end of the web members.

Bibliography

1. *SHS Welding*, British Steel, Corby.
2. *Design of SHS Welded Joints*, British Steel, Corby.
3. Whitefield, S., and Morris, C., 2009, Welded joints with structural hollow sections, *The Structural Engineer*, November 3, London.
4. Hicks, J., 1999, *Welded Joint Design*, 3rd Edition, Industrial Press, New York.
5. Hicks, J., 2001, *Welded Design—Theory and Practice*, Abington Publishing, Cambridge.

15

Orthotropic Floor System

ABSTRACT The chapter begins with the basic principles of orthotropic floor system and continues to discuss the main advantages of this system, particularly in respect of modern steel bridges. Structural behavior and methods of analysis are deliberated, followed by typical details of the different constituent elements in the system with supporting figures. The problem of distortion often faced during fabrication and the remedial measures to avoid this problem have been discussed. The chapter ends with a brief note on corrosion protection of the system.

15.1 Introduction

Welded orthotropic steel floor system with its outstanding structural properties is one of the most important elements in modern steel bridge construction.

The word *orthotropic* is derived from the expressions orthogonal for *ortho* and anisotropic for *tropic*, meaning that an orthotropic deck has dissimilar elastic properties in the two mutually perpendicular directions.

A simple arrangement of an orthotropic steel deck in a bridge is illustrated in Figure 15.1. In this case, the steel deck plate serves as the top flange of three structural members, namely, the longitudinal stiffening ribs, the transverse cross girders, and the longitudinal main girders. Thus, the stiffened deck plate behaves as an integral part of the three supporting members, which collectively form the primary member of the bridge (i.e., the deck), having three separate sectional properties. These are bending resistance in the direction of the longitudinal axis of the bridge, bending resistance in the direction of the transverse axis of the bridge, and torsional resistance about the longitudinal axis of the bridge. Because of the flexibility of the deck with the longitudinal ribs acting as beams on elastic supports, any concentrated load placed on the deck plate is effectively distributed over a wide area to several adjacent cross girders.

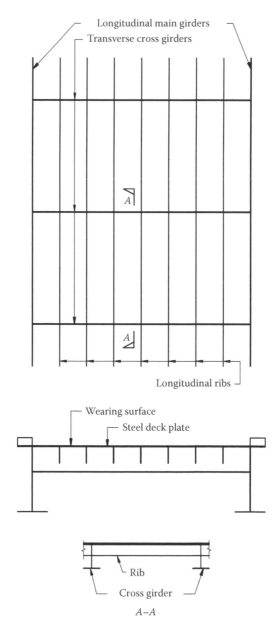

FIGURE 15.1
Typical arrangement of an orthotropic steel deck bridge.

15.2 Advantages

Some of the important advantages of orthotropic steel deck system are discussed in the following paragraphs.

15.2.1 Savings in Weight of the Structure

A light-weight and thin-wearing surface is normally placed on the deck plate, thereby eliminating the traditional heavy concrete floor. This saving of weight, as well as the efficient use of materials, makes orthotropic steel deck system immensely attractive.

Studies in some European bridges of conventional design, which were destroyed in World War II and replaced by new structures using orthotropic steel bridge decks, showed considerable savings in weight per unit area of the deck.

Orthotropic deck system has also been used for reducing the dead load of concrete decks of many existing bridges, thereby effectively increasing their live load carrying capacity. Examples are George Washington Bridge, New York; Throgs Neck Bridge approach, New York; and Benjamin Franklin Bridge, Philadelphia.

In the case of suspension bridges, the bulk of the stresses in the cables and towers is contributed by the dead load of the superstructure. Consequently, reduction of the weight using orthotropic steel deck system instead of traditional system would make the superstructure substantially lighter, entailing reduction in the cost of material.

15.2.2 Reduction in Seismic Forces

For design of structures in earthquake-prone areas, mass is an important factor. The lower the mass, the lower would be the seismic forces. Use of orthotropic plate deck system for structures for such areas becomes useful by reducing the effects of seismic forces. The reinforced concrete deck of the Golden Gate Bridge in San Francisco built in 1937 was replaced by an orthotropic deck system in 1985, thereby reducing the seismic forces in the towers and other bridge components, besides increasing the live load capacity of the bridge considerably.

15.2.3 Saving in Substructure

Saving in dead weight of the superstructure using orthotropic deck system has a direct impact on the substructure design, such as reduction in the footing sizes or in the number of piles. In cases of bridges, the span lengths can be increased, thereby reducing the number of piers.

15.2.4 Ease of Erection

Orthotropic plate deck can be prefabricated in the workshops in large modular units of suitable sizes. These can then be transported to site, and erected with ease even in adverse weather conditions. This system also eliminates the traditional concreting operation after completion of the steel framework, thereby reducing the erection period considerably.

15.2.5 Saving due to Reduction of the Depth of the Structure

Compared to the conventional reinforced concrete deck structure, orthotropic steel bridge deck system offers a thinner deck structure, thereby reducing the construction depth considerably. For this reason, in spite of its higher cost, orthotropic steel deck system may be cheaper for even small span railway bridges, particularly in urban areas, where the cost of construction of the approach embankments is quite high.

15.3 Structural Behavior

In a traditional deck system of a bridge, individual structural components, namely, the deck, the stringers, the cross girders, and the main girders are designed to perform separate and clearly defined functions. Thus, the deck directly supports the wheel loads and transmits these to the stringers. The stringers, which are supported by transverse cross girders, react on them. These cross girders in turn, transmit the loads to the main girders. Thus, the individual components act independently and do not contribute to the strength or rigidity of the other members. Transverse stability is normally provided by lateral bracing systems. In an orthotropic steel deck bridge, on the other hand, the functions of the structural components are intended to work together and closely interrelated. The deck, the stringers, and the cross girders are in effect integrated as one structural element with the steel deck plate acting as a *common* top flange. This stiffened deck, along with the longitudinal ribs, becomes a part of the main girders as their top flange. The deck system also provides adequate transverse rigidity to the main girders, thus eliminating the necessity of separate lateral bracing system. Additionally, the closely spaced grid structure with steel plate deck has a good load distributing capacity of concentrated wheel loads to adjoining panels of the deck system. Consequently, safety against failure in this system is considerably more than that of a conventional bridge floor. In fact, a local load causes an elastic and, eventual plastic stress redistribution to the adjoining elements. This condition eliminates immediate failure of the overloaded element. In case the load is further increased beyond critical limits, the eventual failure would be a local one, and therefore, the structure as a whole would not be affected.

15.4 Analysis

In any element of a loaded orthotropic steel deck, the stresses are due to combined effect of the various functions that the deck is called upon to perform. These stresses are assumed to result from bending of four types of members (see Figure 15.2).

Member I is the deck plate supported by the welded longitudinal ribs and directly supports the load placed between the ribs and transmits the reactions to the ribs. This plate acts locally as a continuous member. Figure 15.2a illustrates such a member.

Member II comprises the plate and the longitudinal rib spanning between cross girders. This member is normally continuous for at least two spans between three cross girders (see Figure 15.2b).

Member III consists of the stiffened deck plate and the transverse cross girders between the two main girders (see Figure 15.2c).

Member IV comprises the stiffened deck plate and the two main girders spanning between the supports and is normally considered to be continuous for at least two spans (see Figure 15.2d). For the purpose of computation of stresses, the effective cross-sectional area of the deck plate along with the

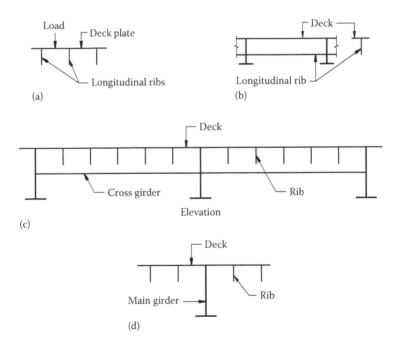

FIGURE 15.2
Types of members in analysis of orthotropic plates: (a) member I—deck plate, (b) member II—longitudinal rib, (c) member III—cross girder, and (d) member IV—main girder.

cross-sectional area of the longitudinal rib is considered as the top flange of the main girders.

In this manner, the deck plate acts as the common top flange for all three members, namely, the longitudinal ribs, the cross girders, and the main longitudinal girders, thereby making the system to behave in an integrated manner. Detailed treatment on the subject is not within the scope of the present text, for which relevant literature and governing codes should be consulted.

15.5 Typical Details

Typical welded details of the main components of orthotropic plate deck system are discussed in the following paragraphs.

15.5.1 Longitudinal Ribs

In general, two basic types of longitudinal ribs are used, namely, open-type and closed-type. Figure 15.3 shows some typical longitudinal rib configurations. The ribs are generally spaced at about 300 mm centers.

Open-type ribs comprise flats, inverted T-sections, angles, bulb sections, and welded to the deck plate. Closed-type ribs consist of trapezoidal, rounded, triangular, and combined shapes. Bent or rolled pieces of steel are welded to the deck plate to form closed-type ribs. Triangular closed-type ribs can be formed by rotating an angle (rolled section or folded plate) by 45° and welding both the legs to the plate. Additional inverted T-section welded at the tip of the triangular closed-type rib increases the flexural strength of the triangular rib considerably. Closed-type rib deck system is essentially a

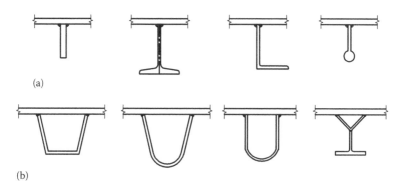

FIGURE 15.3
Typical rib configurations: (a) open- and (b) closed-type ribs.

series of miniature box girders placed side by side with significant torsional rigidity and contributes substantially in distributing the load transversely. On the other hand, torsional rigidity of the open rib deck system is comparatively very small and thus has relatively small capacity for distribution of the load in the transverse direction. It also requires closer cross girder spacing. These factors make the deck system with closed-type ribs more attractive economically, being lighter than a deck system with open-type ribs. Also, in most cases, the quantity of welding required for an open-rib system is significantly more than that in a closed-type rib system. Furthermore, the closed-type rib system has only half the rib surface to protect from corrosion, which reduces the maintenance cost considerably.

The advantage of the open-type rib system is, however, its relatively simple fabrication procedure, requiring only normal degree of precision, and ease in site splicing. The bottom of the open-type rib system is also easily accessible for fillet welding from both sides as also for subsequent inspection and maintenance. Also, transverse bending stresses between the rib and the deck plate is much less compared to that in a closed-rib system and consequently is less prone to fatigue effect.

Typically, for ribs of trapezoidal cross-section, there are two common details for connecting these to the deck plate. These are illustrated in Figure 15.4. The edges of the ribs can either be bevel cut to match the slope of the web, or cut square, as shown in Figure 15.4a and b, respectively. The welding may be done using automatic submerged arc welding method.

Generally, the ribs can be formed to the required trapezoidal shape by means of brake press. Also, these can be obtained by rolling process. One advantage of the trapezoidal ribs is that these can be nested one above the other, and thus can be stored and transported economically.

15.5.2 Transverse Cross Girders

Transverse cross girders may comprise inverted T-sections (rolled T-section, T-sections cut from rolled I-sections, or built-up T-sections) using the deck

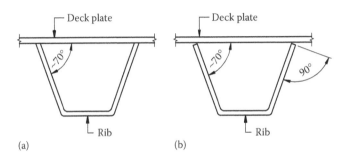

(a)   (b)

FIGURE 15.4
Edge details of ribs: (a) bevel square cut edges.

plate as the top flange. Alternatively, rolled shapes (I-beams, channels, etc.) or full-depth diaphragm plates (in box girders) welded to deck plate can also be used.

15.5.3 Splices of Longitudinal Ribs

Typically splices of longitudinal ribs can be detailed in two ways:

1. The ribs are discontinued at the transverse cross girders and joined preferably by single bevel butt weld to the web of the cross girders. The bending stresses in the ribs are transferred through the web of each cross girder by means of the welds. With this detail, however, there is a risk of lamellar separation in the web of the cross girder, because of transverse force applied through it. Figure 15.5 illustrates typical splice details for ribs discontinuous at cross girders.

2. The ribs are made continuous through slots in the web of the cross girders to suit the profile of the ribs (Figure 15.6). The splices in the ribs may be located at every 12–18 m, depending on the length

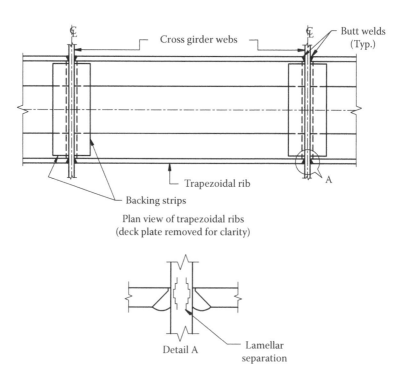

FIGURE 15.5
Typical splice details for ribs discontinuous at cross girders.

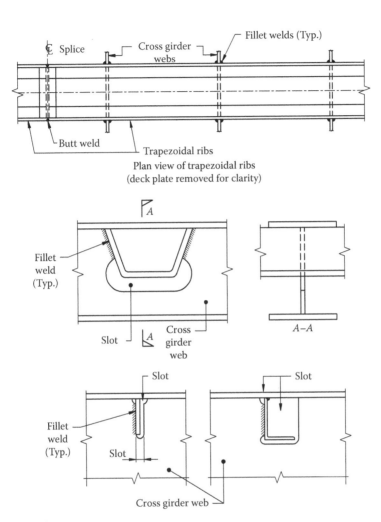

FIGURE 15.6
Typical splice details for ribs continuous at cross girders.

available, and away from the locations of the cross girders. With this detail, the risk of lamellar tearing in the webs of the cross girders is eliminated. This detail is thus preferred, particularly in the portions of the deck subjected to tension. Also, the quantity of welding is reduced considerably.

15.5.4 Site Splices of Panels

It is a common practice to prefabricate the deck in panels of suitable sizes in the shops, transport these panels to the site, erect, and then connect these by welding or bolting at site. The size of the prefabricated panels will depend on transportable size and/or the facilities available at site. This arrangement offers maximum use of automatic downhand welding and efficient fabrication methods, including use of jigs for the production of deck sections, thereby resulting in saving of time as well as cost. Although bolted site connections are easy to perform and do not require any particularly expert workmen at site, the protruding splice plate and bolt heads make the wearing surface at splice locations thinner. Also, loss of cross-sectional area due to bolt holes in the tension area reduces the strength of the member to some extent. In welded site connections, these problems can be eliminated. Consequently, these are most commonly used nowadays. However, they need considerable degree of precision to avoid distortion of the plates due to transverse and longitudinal shrinkage of welds. Distortion depends largely on structural details, size of welds, welding methods, and sequence of welding, and should be assessed correctly and appropriate remedial measures taken.

Longitudinal trapezoidal ribs are normally connected by butt welding from outside, using a backing strip placed on the inside profile of the trapezoid. In order to avoid problem in the fit up of the two (joining) ribs, ends of the ribs may be left unwelded to the deck plate for about 300 mm on each side of the joint. These welds are to be completed after welding of the rib splice joints. This arrangement will allow easier horizontal alignment of the joining ribs for the purpose of butt welding. Also, to provide accessibility for site welding of the rib splice, the deck plate may be kept short by about 250 mm from each end of the ribs. After the site welding of the rib splice is over, a 500 mm wide deck plate would be inserted to fill the gap and welded to the main deck plates by two transverse butt welds. The ends of the ribs are to be welded (overhead) to the 500 mm wide deck insert plate. Figure 15.7 illustrates the arrangement. An alternative method of joining the trapezoidal ribs is to butt weld these from inside, placing the backing strip on the outside profile of the rib. In this case, welding can be done while working from the top.

15.6 Distortion

Fabrication of orthotropic deck plate involves a considerable amount of rib-to-web welding, and that too on one side of the plate only. As a result, the plate is generally susceptible to distortion.

Distortion affects not only the riding quality of the deck but also the structural stability of the compression flange (e.g., in box girders). Furthermore, for site welded splices, variation in the geometry of the individual panels needs to be within the tolerance limits.

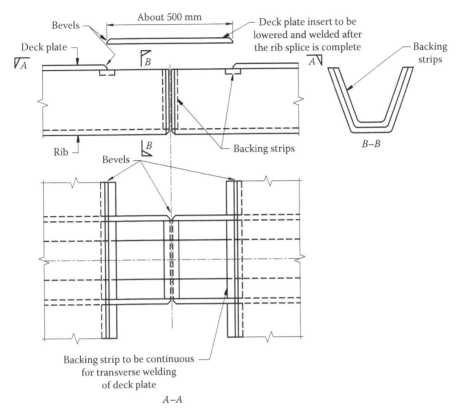

FIGURE 15.7
Arrangement of site splice of trapezoidal ribs.

Broadly, there are two options to ensure that the geometry of the finished panel is free from distortion. These are as follows:

- Prevention prior to welding
- Correction after welding

Prevention involves prebending of the plate by clamping it to a curved assembly bed, so that the distortion is effectively offset after welding. The alternative method of correction after welding is generally done by heat treatment to straighten the distorted panel. In practice, a combination of the two methods is often used. However, both the methods require considerable experience and practice to produce an acceptable final product. The topic of distortion has been discussed in detail in Chapter 4 to which reference may be made.

15.7 Corrosion Protection

In an orthotropic deck system, two locations, which are inaccessible after erection, deserve special attention. These are inside of closed ribs and top surface of the deck plate. The other surfaces, which are accessible, can be painted for corrosion protection.

Closed ribs are normally made air tight during fabrication. Since atmospheric corrosion cannot occur without the presence of air, once the inside of the closed rib is made air tight, the necessity to make any provision for corrosion protection becomes minimal. However, when bolts are used for field splicing the closed ribs, the hand holes required for bolting make the inside vulnerable to corrosion. In such cases, air tightness must be ensured by welding diaphragm plates inside the rib section on either side of the splice.

A bituminous-wearing coat over the steel deck plate does not provide any satisfactory protection against corrosion. It is therefore necessary that the bond coat, which is first applied on the top surface of the steel deck, serves as the protective layer against corrosion.

Bibliography

1. Wolchuck, R., 1963, *Design Manual for Orthotropic Steel Plate Deck Bridges*, American Institute of Steel Construction, New York.
2. Blodget, O.W., 2002, *Design of Welded Structures*, The James F. Lincoln Arc Welding Foundation, Cleveland, OH.
3. Xanthakos, P.P., 1994, *Theory and Design of Bridges*, John Wiley & Sons, New York.
4. Mangus, A.R., and Sun S., 1999, Orthotropic deck bridges, in: *Bridge Engineering Handbook*, Eds., Chen, W.F., and Duan, L., CRC Press, New York.
5. Stainsby, D., 1994, Floors and orthotropic decks, in: *Steel Designers' Manual*, Eds., Owens, G.W., and Knowles, P.R., Blackwell Scientific Publications, Oxford.
6. Ghosh, U.K., 2006, *Design and Construction of Steel Bridges*, Taylor & Francis Group, London.

16

Economy in Welded Steelwork

ABSTRACT This concluding chapter opens with a brief discussion of the mechanics of costing. Various factors that affect welding costs in the design and fabrication stages have been deliberated, followed by discussions on general topics, namely, overhead expenses and labor costs.

16.1 Introduction

Three basic considerations must be borne in mind when designing and fabricating a welded steel structure or component. These are as follows:

1. Efficiency
2. Aesthetics
3. Cost

Obviously, the designer's prime consideration would be to design the structure to be strong enough to carry the loads imposed on it. The next consideration would be to ensure that the completed structure has an attractive appearance with simple and clean lines, a prime requirement of contemporary architects. It is in this area that welding takes a lead to be the most appropriate method of fabrication.

The third consideration, namely, the cost, is very important to both the customer and the fabricator and has considerable impact on the popularity of welded structures globally. The various factors that affect the cost of welded structures are briefly discussed in the present chapter.

16.2 Mechanics of Costing

As in the case of any other operation, the costs of welding basically come under two heads, namely, direct costs and indirect costs. Costs of labor and consumables are grouped under direct costs, while overhead costs come typically under indirect costs. An analysis based on these costs will provide a very easy-to-apply system.

16.2.1 Direct Costs

16.2.1.1 Labor Cost

A welder is generally associated with only one job. Therefore, it should be easy to identify and record the time (and consequently the costs) for labor element in the particular welded job.

Labor cost is a function of time and therefore should be the time spent on actual welding, that is, the actual time that the arc is burning. This time is termed *arcing time*. In effect, however, the welder is paid for all the time between his reporting for duty and the finish of the shift. The time spent for *non-arcing* operations is also considered as spent on the job, and is charged as labor cost on the job. These *non-arcing* operations include setting up the job, de-slagging, and cleaning the welds, changing electrodes, and aligning the joint. Apart from these tasks, the welder may have to wait for the work to be delivered, discuss about problems faced during welding operations, and draw consumables/electrodes from the stores. These non-arcing times are included in the welder's chargeable time for the particular job. In order to put all these time elements into reckoning, it is necessary to express arcing time as a proportion of the total time charged to the job, which is termed *operating factor* and is expressed by the ratio:

$$\text{Operating factor} = \frac{\text{Arcing time}}{\text{Total time}} \tag{16.1}$$

This factor thus indicates the proportion of the welder's time, which he uses for depositing the weld metal and depends largely on the conditions of the shop, the type of work involved, and the help given to the welder in the form of helper (unskilled worker). The increased cost for this additional unskilled worker has to be taken into account while assessing the total welding cost.

Apart from welding cost, other costs such as cost of cutting the components to required size and preparation of edges (for butt welding) are also to be included for direct labor cost.

16.2.1.2 Costs of Consumables

For manual metal arc welding the quantity (and hence the costs) of consumables can be assessed by calculating the volume of weld metal from the geometry of the joint, making allowances for wastage due to stub ends, spatter, or incomplete use of electrode due to access difficulty.

For automatic welding, the length of continuous wires and quantities of flux can be obtained from the manufacturer's data tables, making allowances for wastage. Cost of electricity, shielding gas, and so on can be determined once the arcing times have been established.

16.2.2 Indirect Costs

Overhead costs come under indirect costs. These include a share of centralized expenditure, which are additional to labor and consumables costs discussed in the previous paragraphs. The centralized expenditure comes under two heads, namely, workshop and general overheads.

1. Workshop overhead costs include the following:
   a. Rent or depreciation and taxes for the workshop
   b. Charges for electricity, water, and so on
   c. Depreciation for equipment
   d. Salaries and wages
   e. Repairs and maintenance
   f. Inspection and quality control
   g. Planning and monitoring
   h. Stores
   i. Internal transport
   j. Canteen, first aid, and so on
2. General overhead costs include the following:
   a. Office rent (or depreciation) and taxes
   b. Air conditioning and lighting
   c. Transport
   d. Salaries
   e. Marketing expenses

One of the most common methods of assessing overhead costs is to add a fixed percentage to the labor costs. This percentage varies considerably, depending on the types of the organization and the structures being fabricated.

16.3 Factors Affecting Welding Costs

Many factors affect the overall cost of a welded structure. These should be examined both in the design office and in the workshop level. Often, it has been found that when the fabrication drawing reaches the shop floor, it is too late to improve the uneconomical designs, as these have already been approved by the client and materials ordered or a part already delivered. Thus, the responsibility of producing an economic welded structure rests not only on the fabricator at the workshop level but also on the designer in the design office. In view of this aspect, it is proposed to discuss the ways and means of reducing costs in two levels, namely, first in the design office starting from the concept stage of the design process up to detailing when the ideas for improving the design can be incorporated in the design and drawing and then in the fabrication shop where practical aspects of saving can be implemented. It is recommended that the design office and fabrication shop should exchange their ideas prior to finalizing the design/detail of the welded component/structure.

16.3.1 Design Stage

16.3.1.1 Choice of Sections

By using arc welding, it is possible to fabricate a wide range of structural forms and shapes. Thus, for a particular component of a structure, it is possible to produce an ideal structural form by gas cutting plates and welding these to form the required shape. However, it may also be possible to use a comparatively heavier *standard section* (though less suitable theoretically), which may produce a safe, and at the same time a cheaper job. Thus, it is necessary, for the sake of economy, to make a compromise between the ideal section and a standard section, although the former may be lighter in weight. A word of caution in this connection is pertinent. Cheapness should *not* be the only consideration, and alternative designs need to be weighed against each other, giving due consideration to design requirements, as well as to costs. An example will illustrate a possible situation. An ideal *I* section may be built up by two flange plates and a web plate by providing four lines of fillet welds at the connection locations (see Figure 16.1a). Assuming that the section is required to act as a column, a section having the same strength may be obtained by placing two rolled steel channels or beams side by side and providing two rows of welds (see Figure 16.1b through d). In order to decide on the section to be used, consideration must be given to the costs of material (plate and sections) and welding. The other factor that needs consideration is satisfactory welding of the channel and beam sections. Deep penetration welds may provide satisfactory welding; see Figure 16.1b through d. Alternatively, butt welds with prepared edges may be a preferred solution.

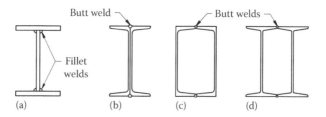

FIGURE 16.1
Built-up column sections: (a) section built-up with plates and (b–d) sections built-up with rolled channels and beams (b) through (d).

One other aspect needs consideration. The rounded toes of the channels and beams may have to be built up by additional runs of weld metal to obtain the necessary throat thickness or these can be machined square. In the latter case, reduction of effective area also needs to be considered.

16.3.1.2 Welding Position

It has already been discussed (Chapter 7) that flat (downhand) position is the most advantageous for welding due to the following reasons:

- Maximum deposit rate of weld metal
- Maximum speed of welding
- Least cost of welding
- Ease of welding and least fatigue of welders

The advantage of the flat position of welding should be particularly considered while designing joints, which need to be site welded and which cannot be rotated by a manipulator (see Chapter 7, Figure 7.8) or turned over by a crane. Thus, a butt weld to be made at site in the flanges of a girder could be detailed either as a single-V or a double-V preparation. Although in a single-V preparation, more weld metal will be required (than in a double-V preparation), the welding would be in flat (downhand) position. In this case, for overall economy, a single-V butt weld will be preferable. Alternatively, a butt weld with asymmetrical double-V preparation as shown in Figure 11.12 (Chapter 11) may be used.

Another case of design detail using flat position of welding is illustrated in Figure 16.2. In this example, an existing beam is required to be strengthened by addition of flange plates. If it was a new construction, where the beam could be rotated, the common detail would have been to use two additional

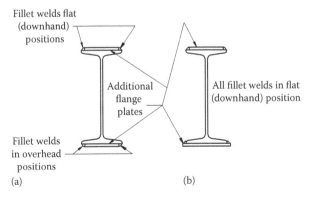

FIGURE 16.2
Strengthening of existing beam.

plates, somewhat narrower than the flange. However, since the beam is already in its place and cannot be moved, the detail as shown in Figure 16.2a would have meant overhead fillet welds for connecting the lower additional flange plate. Therefore, the lower flange has been made wider to eliminate overhead welding and all the welds are to be made in the flat position as shown in Figure 16.2b.

16.3.1.3 Accessibility of Welds

It is the responsibility of the designer to ensure that all welds are easily accessible to the welder. Often, due to oversight in the design stage, welds indicated on a drawing are difficult or impossible to make. Some typical details with lack of access for welding are shown in Figure 7.6 (Chapter 7). Such solutions should be avoided by carefully checking this aspect before finalizing a joint detail. Otherwise, the lapse may be very expensive to correct at the fabrication stage.

16.3.1.4 Joint Preparation and Weld Volume

In a butt weld, cost of welding largely depends on the volume of the deposited weld metal, which varies according to the type of edge preparation adopted. Thus, a single-V butt weld requires more weld metal than a double-V butt weld. Also, for plates up to 20 mm thickness, a single-U butt weld requires more weld metal than a single-V butt weld. Similarly, for plates up to about 35 mm thickness, a double-U butt weld requires more weld metal than a double-V butt weld. However, for overall economic evaluation, it will be necessary to take into consideration other factors also. Cost of edge preparation is one such factor. Single- or double-V preparations can be easily done by flame cutting, while single- and double-U preparations need careful machining. Thus, U-preparations are slower as also costlier operation. Also, because of the steeper sides, the possibility of under cutting and slag entrapment is greater in a butt weld using U-preparation. Therefore, it is important that the type of butt weld should be carefully selected to ensure that the most economical shape, consistent with efficient welded joint, is obtained.

In case of a fillet weld, the strength of the joint is proportional to the product of the throat thickness and the length of the weld. Therefore, the strength of a fillet weld is directly proportional to the throat thickness (i.e., size) or the length. On the other hand, the volume of the weld metal, which is a measure of the cost, is proportional directly to the length and to the square of the throat thickness. Therefore, if the length is doubled, the strength, as well as the cost, is doubled. However, doubling the size of the weld, doubles the strength, but increases the cost four times. For economy, therefore, the size of a fillet weld should be kept as small as possible. It is also more economical to use continuous fillet weld of small size than intermittent welds of a larger size. However, as per national codes, the minimum size of fillet weld is governed by the thickness of the material being welded. Therefore, where the required size of continuous weld is limited by codal provisions, intermittent weld of the minimum size allowed may show some saving in the cost.

16.3.2 Fabrication Stage

16.3.2.1 Rectification of Mistakes

A major cause of cost overrun in welded fabrication work is rectification of mistakes. These often entail not only a repeat of the fabrication and welding costs but also additional costs for dismantling the faulty work, usually by laborious chipping. Mistakes may be due to various reasons namely, misinterpretation of drawings/welding procedure, use of wrong type of electrode, faulty edge preparation, and so on. It is therefore necessary that the coordination between various units involved in the fabrication, including the design/drawing office, should be very strong. Proper link between the welding shop and the design/drawing unit is particularly necessary to avoid/minimize mistakes in the fabrication work.

16.3.2.2 Accuracy of Edge Preparation and Fit-Up

Accuracy of edge preparation is very important for welding work, particularly if the automatic arc welding process is used. For proper edge preparation, it is imperative that the welding shop must have suitable equipment for preparing plate edges, such as flame cutting and planing equipment. Inaccurate edge preparation would result in unacceptable welded joint, which would have to be rectified or redone, entailing additional expenses.

16.3.2.3 Jigs and Manipulators

For repetitive work, jigs and fixtures should be used wherever possible. These will increase accuracy considerably and reduce setting up time. Manipulators (Figure 7.8) are also used in the setting up, but they are primarily used for positioning the job, so that majority of the work can be done in the flat (downhand) position. Manipulators are very useful equipment for economy in welding. However, their usefulness should offset their initial cost and running expenses to warrant their installation in the welding shop.

16.3.2.4 Choice of Welding Process

Cost of welding can largely be influenced by the type of welding process being used. Apart from the difference in the initial installation cost, there is considerable variation in the running cost as well for different processes.

It is therefore necessary to consider this aspect while computing the overall cost of welding for a particular job.

For automatic welding process, it is imperative that the work should be designed and detailed to ensure that the welds are in position within easy reach of the machine and the welding head can be properly positioned. Also, the run should be long enough to achieve a sufficient saving in time to warrant the increased setting up time of the machine. On the other hand, in the case of semiautomatic machines, because of their portability, the machines can be conveniently brought to the job, and shorter runs become economical also.

16.3.3 General Remarks

16.3.3.1 Overheads

As already discussed, overhead costs include a number of centralized expenditure in addition to costs due labor and consumables. These items, though a function of the management, also contribute to the control of welding costs. A good working condition in the fabrication shops, comprising of good lighting, fume extraction system, and heating where necessary, should be provided. Also, there should be an even balance of laborers and inspectors in relation to the number of welders. Collectively, these should improve the quality, as well as output, of the work.

16.3.3.2 Labor Costs

Labor cost depends on the welding time. Therefore, it is important to reduce the total working time by careful planning of production, which will minimize delay or hold up in the flow of work and help in enhancing the output. In order to achieve this, every effort should be made to maintain a high operating factor.

It is often found that the welder is welding for only a fraction of his total time, the remaining time being spent in deslagging the welds, fetching electrodes, setting up the job, and so on. It will pay to provide the welder an *assistant* (with lower pay) to carry on the ancillary jobs and leave the welder to weld, thereby improving the operating factor.

Welders may also waste valuable time while waiting for a crane. It would perhaps be economical for a work to be carried out in a welding shop area equipped with a dedicated crane for welding work only.

In certain cases, it may be advantageous to split the job into a number of subassemblies, employing separate welders for each subassembly to be done in parallel. The final assembly consists of joining these subassemblies. This method not only speeds up the production cycle but also ensures better quality control of each subassembly, prior to the final assembly.

16.4 Concluding Remarks

In the preceding sections, an attempt has been made to provide a broad picture of the different factors, which influence the cost of arc welding. It is hoped that this chapter will help the designer/detailer and personnel in workshop level in acquiring some basic knowledge on the subject and encourage them to appreciate their importance in producing not only a functionally sound but also an economical welded structure.

Bibliography

1. Brooksbank, M.A., and Waller, W.D, *Economics of Arc Welding*, Quasi-Arc, Bilston.
2. Smith, J.L., 1966, Economies in shop and site welded structural steelwork, *British Welding Journal*, December, pp. 699–706.
3. Reynolds, D.E.H., 1963, Effect of Economics on the Design of Joints in Welded Fabrications, *British Welding Journal*, October, pp. 498–507.
4. Dowling, P.J., Knowles, P.R., and Owens, G.W. (eds.), 1988, *Structural Steel Design*, The Steel Construction Institute, London.
5. Gourd, L.M., 1995, *Principles of Welding Technology*, Edward Arnold, London.
6. *Procedure Handbook of Arc Welding: Design and Practice*, 1959, The Lincoln Electric Company, Cleveland, OH.

Index